Facts for General Agriculture

NIPA® GENX ELECTRONIC RESOURCES & SOLUTIONS P. LTD.
New Delhi-110 034

About the Authors

Ms. Kawaljeet Kaur is working as an Assistant Professor, Department of Agriculture, CT Group of Institutions, Jalandhar, Punjab, India. Since 2019 teaching undergraduate students. She obtained B.Sc. Hons. Agriculture (2017) with first class and completed M.Sc. Soil Science and agriculture chemistry in the year 2019 with first class as well from lovely professional university. Also she got young scientist award and young soil scientist award in the year 2021. She has attended National and International level conference and also attended National and International webinars, workshops, and online quiz, certification programmers. She has published many research papers and review papers in reputed journals, popular articles, book chapters and many books.

Mr. Anupam Singh is Ph.D scholar (Horticulture) Department of Agriculture, IIAST, Integral University, Lucknow. He has completed B.Sc. Agriculture (2014) with first division from VBS Purvanchal University, Jaunpur and completed M.Sc. (Ag) Horticulture in the year of 2017 with first division from ANDUA&T Ayodhya. Mr. Singh is qualified ICAR - NET in Fruit Science in 2018. He has attended more than 10 National and International level conference and also attended more than 12 National and International webinars, workshop and training. He has published research papers and review papers in reputed journals and popular articles Hindi and English both and book chapters also.Mr. Singh has got Young Scientist Award 2019 by ATDS society Ghaziabad.

Dr. Hina Upadhyay is an Associate Professor at Lovely Professional University in Punjab, India, where she works in the Department of Agronomy, School of Agriculture. She was formerly an Assistant Professor in the Department of Life Sciences at Boston College and Jiwaji University in Gwalior. From 1996 to 1998, she worked as a project fellow at Jiwaji University's Aerobiology Laboratory, S.O.S Botany. Among other scientific societies and organizations, she is a member of the Association of Microbiologists of India, a Fellow of the Society of Bionaturalists, a Life Member of the Indian Botanical Society, and a Life Member of the Indian Science Congress. At a variety of international and national seminars, symposia, and conferences, Dr. Hina chaired technical sessions and presented papers. In addition, from 2010 to 2013, she was a member of the Board of Studies in Botany at Jiwaji University in Gwalior, and from 2019 to 2020, she was a sectional member of the Indian Science Congress's Agricultural Section. She has published 45 peer-reviewed research and review papers, as well as 7 book chapters with Springer International Publishing and Elsevier Science Publishing. Her current research area includes agroecosystem sustainability.

Dr. Subhash Chand presently working as a Professor in the Division of Soil Science, Sher-E-Kashmir University of Agriculture Sciences and Technology of Kashmir, Shalimar Campus, Srinagar, UT of Jammu and Kashmir. He has 17 years experience of research, teaching and extension. Prof. (Dr.) Subhash Chand, Co-chaired two technical session at International Conferences at Jaipur and Pantnagar. He is member of several scientific societies. He is elected member of Association of Soil and Water Conservationist, Dehradun for 22-23. He was councillor for Indian Society of soil survey and Land use planning. Nagpur for 2014-15. Dr. Chand presented paper at Humbolt University, Berlin, Germany in 2010. He published research and review paper in National and International Journal of repute.

Facts for General Agriculture

Kawaljeet Kaur
Assistant Professor
Department of Agriculture
CT Group of Institutions
Jalandhar, Punjab, India

Anupam Singh
Ph.D Scholar (Horticulture)
Department of Agriculture
IIAST, Integral University
Lucknow, Uttar Pradesh, India

Hina Upadhyay
Associate Professor
Lovely Professional University
Jalandhar, Punjab, India

Subhash Chand
Professor
Division of Soil Science
Sher-E-Kashmir University of Agriculture Sciences and Technology of Kashmir
Shalimar Campus
Srinagar, UT of Jammu and Kashmir, India

NIPA® GENX ELECTRONIC RESOURCES & SOLUTIONS P. LTD.
New Delhi-110 034

NIPA® GENX ELECTRONIC
RESOURCES & SOLUTIONS P. LTD.

101,103, Vikas Surya Plaza, CU Block
L.S.C. Market, Pitam Pura, New Delhi-110 034
Ph : +91 11 27341616, 27341717, 27341718
E-mail: newindiapublishingagency@gmail.com
Web: www.nipabooks.com

For customer assistance, please contact
Phone: + 91-11-27 34 17 17
Fax: + 91-11- 27 34 16 16
E-Mail: feedbacks@nipabooks.com

ISBN: 978-93-95763-04-2

Composed and Designed by NIPA®.

Preface

This book is a result of a thorough and sincere analysis of test questions from the ASRB, ARS, SRF, JRF/NET, and numerous agricultural universities, in addition to bank exams. To demonstrate it takes work to compile. Modern inquiries that is systematic in their reliability and usability. Each applicant will receive a sample question from this set of questions, which will help them be more able to concentrate. The main objective of this book is to let the readers fully understand the information from numerous several agricultural disciplines, including agronomy, soil science, Horticulture, Entomology and Agriculture terminology etc. But only if this book satisfies everyone's needs I should consider my work to have been worthwhile if I am participating. Readers are invited to flag out any errors or omissions and offer their insightful suggestions for how to make the book better. Authors are hopeful that book will be extremely helpful in agriculture competitive exams.

Authors

Contents

1

Important Facts of Agriculture

1. Father of agronomy is - **Peter Decresenzi**.
2. **Dr. Biwant Rajis** the founder of Relay cropping.
3. The Indian Institute of Farming System Research is located at - **Modipuram, Meerut, U.P.**
4. Scientific study of mixed cropping was firstly done by - **La-Flitze (1928)**.
5. Relay cropping concept was derived from - **Relay race**.
6. Natural Resource Management is an example of **Sustainable Agriculture**.
7. Cropping System is a component of - **Farming System**.
8. When the sub crop are sown to supplement of yield of the main crop , the sub crops are known as **- Augmenting Crops**.
9. The main crop is grown in the centre, surrounded by hardly or thorny crops, the sub crop is known as **- Guard Crops**.
10. The yield of both crops, grown together are found to be higher than the yield of their pure crops on unit area, such cropping is known as - **Synergetic Cropping**.
11. Cash Crops are those crops, which may be sold directly from the field without processing like vegetables, potato, etc.
12. Formula of Cropping Intensity = **Total Cropping Area /Total Sown Area Multiple by 100**.
13. The growing of two or more crops on the same field in a year is called - **Multiple Cropping**.
14. The practice of raising grasses or pasture, is also known as - **Ley Farming**.
15. In India, the maximum acreage and production of wheat is in **U.P.**
16. The type of fruit of wheat is **Caryopsis**.
17. The optimum temperature for ideal germination of wheat is **20 to 25°C**.

18. Norin-10 gene was brought to USA by - **S.C. Salamone 1984**.
19. The dwarf winter variety of wheat first time developed through using Norin gene by - **Dr. O.A.Vogel (USA)**.
20. Triple gene dwarf variety were released in **1970**.
21. Double gene dwarf varieties of wheat are - **HD 2204, Janak, UP 215, Pratap, Kalyansona, Arjun, Sonara 64**.
22. Single gene dwarf varieties are - **Sonalika, Girija, UP-262, Wl 711, Lerma Rojo 64A**.
23. The protein found in wheat flour which is helpful for chapati making - **Gluten**.
24. Central Zig - Zag axis of wheat is called -**Rachis**.
25. Mexican dwarf wheat was evolved by - **Dr. N.E. Borlaug**.
26. The toxic alkaloid found in the leaves of sorghum crops is - **Dhurin or HCN**.
27. CSH 1 was released as first dwarf sorghum hybrid in **1964**.
28. Which crop is known as camel crops - **Sorghum**.
29. Native of Jower is Africa **(Ethiopia & Sudan)**.
30. The maximum area and production of Jowar is in **Maharashtra**.
31. Jowar is poor in lysine but rich in - **Leucine**.
32. The inflorescence of jower is called – **Panicle**.
33. First hybrid variety of bajara is **HB-1 (1965)** developed at PAU, Ludhiana.
34. Protein content in Bajra is **11-12%**.
35. Origin of Bajra is **Africa**.
36. Bajra is sensitive to - **Water Logging and Acidic Soil**.
37. The largest producer state of cotton in India is - **Gujarat**.
38. Indian Institute of Weed Science is located at - **Jabalpur**.
39. In cotton, the maturity of fibre is judged by - **Arealometer**.
40. **H-777** is the cultivar of cotton.
41. Hybrid-4 is the first commercial cotton of the world developed by - **Dr. C.T. Patel in 1970**.
42. The leaves of cotton becomes radish in colour due to decrease **Nutrient Uptake and Moisture**.
43. Origin of cotton is from - **India**.

44. Cotton belongs to - **Malvaceae**.
45. Intra-specific hybrid of cotton is - **Hybrid-6, Savita, Surya**.
46. Inter-specific hybrid of cotton is - **Varalaxmi, DCH-32, HB -224** etc.
47. One seed of cotton has **60000-80000** fibres.
48. Ginning % = **wt. of lint / wt. of seed cotton taken for ginning multiply by 100**.
49. The removal of lower leaves which come in contact with soil and lose their commercial value is known as **Priming Process**.
50. Priming method is used in - **Cigarette and Wrapper Tobacco**.
51. The main aim of Topping and Desuckering is - **to divert the energy and nutrient of plant from flower to leaves**.
52. The removal of the lateral branches or suckers or auxiliary buds is called- **desuckring**.
53. Central Tobacco Research Institute is situated at - **Raja-h-Mundri, A.P. (1947)**.
54. The richest source of protein among the food grain is - **Pulses**.
55. Headquarter of Indian Institute of Pulses Research is located in - **Kanpur U.P.**
56. Which pulse crop doesn't fix N_2 nitrogen from atmosphere -**Rajma**.
57. The word pulse is derived from - **French world pottage or poride means-soup (dal)**.
58. The term legume is taken from - latin word - **means to collect**.
59. India is the largest producer of - **Gram**.
60. The sour taste of gram leaf is due to presence of - **Maleic Acid (60%) and Oxalic Acid**.
61. Shelling % of Pea is **49**.
62. Harvest Index of arhar is **19%**.
63. Origin of aarhar is from - **Africa**.
64. Queen of cereals is called - **Maize**.
65. The composite cultivars of maize are - **Kisan, Jawahar, Amber, Vijay Sona, Vikaram, etc.**
66. Maize is a **C_4 Plant**.
67. Maximum Maize producing state is **U.P.**

68. The maize Protein is called - **Zein**.
69. Male inflorescence of maize is called - **Tassel.**
70. The critical stage for water application in maize is - **Tasseling and Silking stage**.
71. Which Indian scientist shared World Food Prize for miracle maize - **Dr. Sruinder K. Vasal (2000)**.
72. Maize holds in origin from - **Mexico**.
73. All India Coordinated Maize Improvement Project was started in - **1957 at New Delhi**.
74. Double cross technique for hybrid seed production was developed by - **D.F. Jones (1920)**.
75. The idea of hybrid maize was first conceived by : **E.M. East and G.H. Shull (1910)**.
76. The criteria for harvesting of sugarcane - **withering of lower leaves, brix 20%, sucrose 15%**.
77. Algin planting method of sugarcane is evolved by - **Sam Higginbottom University of Agriculture, Technology and Sciences (Deemed to be University) Prayagraj, Uttar Pradesh**
78. Adsali sugarcane is common in **Maharashtra**.
79. Sugarcane is planted in furrows because - **it reduce lodging**.
80. Inflorescence of sugarcane is generally called - **Arrow or Open Panicle.**
81. Sucrose content in cane is - **13-14%**.
82. Sugar from juice is - **6-10%**.
83. Hand refractometer reading for sugarcane maturity is **20**.
84. Sugarcane Breeding Institute is situated at - **Coimbatore**.
85. Adhsali sugarcane crop mature in **18 months**.
86. father of hybrid rice is - **Yuan Long Pin**.
87. India's first dwarf variety released during **1968 was jaya**.
88. World's first high yielding semi dwarf basmati rice variety is - **Pusa Basmati-1**.
89. World's first super rice variety for saline/alkaline soil - **Lunishree**.
90. Harvesting of paddy is done at moisture content - **21-23%**.
91. First mutant variety of rice **is Jagannath** (gamma ray irradiation of T-141).

92. Which gas is released from field of paddy - **Methane** (CH_4).
93. SRI techniquecue was originated in Madagascar in 1987 by Prof. **Henry De Laulanie**.
94. First super fine aromatic basmati hybrid is - **Pusa RH-10**.
95. Oilseed Production Programme (OPP) was started during - **1990-91**.
96. Technology mission on oilseed was started in **1986**.
97. National Research Centre for Groundnut was established at **Junagarh (Gujarat).**
98. Indian Central Oilseed Committee was established **1947**.
99. All India Coordinated Project on Oilseeds was established **during 1967**.
100. Main adulteration in mustard - **argemone mexicana oil**.
101. Mustard fruit is known as - **siliqua**.
102. Pungency of mustard is done to - **isothiocynate**.
103. National seed corporation was established in March - **1963**.
104. Seed Act was passed in **1966**.
105. Tetrazolium Test is done for - **Viability and Vigour**.
106. The Indian Seed Act came into effect **on 2 Oct. 1969**.
107. NSC started functioning in July **1963**.
108. The headquarter of NSC is - **Pusa Campus, New Delhi**.
109. National Seed Project was launched in - **1976**.
110. The world meteorology is derived from - **Greek language**.
111. International Meteorological Organization was established in - **1873**.
112. Division of Agri. Meteorology is at Pune, Maharashtra was established in **1932**.
113. Solar radiation is measured by - **Pyranometer** (total incoming radiations).
114. Agro Ecological Zones in India is 8 (by NARP (ICAR), 21 by NBSSLUP)- **but now only 20**.
115. Agro-Climatic Zones (15 nos.) **are formed in India during 7th five year plan (1988)**.
116. The green house gases are - **CO_2, CFC, CH_4, etc.**
117. Indian Meteorological Dept. was established in the year of - **1875**.
118. Wind velocity is measured by - **Anemometer**.

119. Father of agro-climatology is - **Koppen**.
120. Koppen classified the climate based on - **Temperature & Precipitation**.
121. The study of weather is called - **Meteorology**.
122. The term weed was firstly used by - **Jethrotull**.
123. Indian Institute of Weed Science is situated at - **Jabalpur (M.P.)**.
124. Commercial mycoherbices first appeared in market **USA in early 1980s**.
125. National Horticulture Board was established at **Gurgaon (Haryana)**.
126. First All India Coordinated Floriculture Improvement Project was started in **1971**.
127. Central Temperate Horticulture Research Institute is situated at - **Srinagar U.T. of J&K**.
128. International Institute of Horticulture is situated in - **Brazil**.
129. Who is the father of pomology - **Decandolle**.
130. National Horticulture Board was established in the year **1984**.
131. Who was the first Deputy Director General of Horticulture, **ICAR - Dr. K.L. Chadha**.
132. Which vegetable rank first among vegetable crops in production in the world - **potato**.
133. In India, potato was introduced by **Portuguese in 1965**.
134. Research on potato was stared at IARI New Delhi in - **1935**.
135. The International Potato Centre was established **in 1971 at Lima (Peru)**.
136. Central Potato Research Institute was established in - **1949 (Kufri, H.P.)**.
137. Seed plant technique was developed by - **Dr. Pushkarnath**.
138. The Headquarter of CPRI was Shifted from **Patna to Shimla in 1956**.
139. True potato seed concept was first realize to raise commercial crop in India by **Dr. S.Ramanujam (Fist director of CPRI)**.
140. Cauliflower was introduced in India by **Dr. Jemson from Kew Garden London in the year 1822**.
141. Male sterility in cauliflower was introduced by **Peerson (1981)**.
142. The toxic substance of cucurbits is - **Cucurbitacin**.
143. Which hormones are used in cucurbits to modify sex and to induce femaleness - **GA & NAA**.

144. Seedless variety of mango is - **Sindhu (devloped at FRS vengurula).**
145. Most exported variety of mango is - **Alphonso.**
146. Black tip in mango is a physiological disorder due to **brick clin fumes mostly CO, SO_2 or deficiency of boron.**
147. Which type of incompatibility is found in mango - **Gemtophytic.**
148. Mango variety, most suitable for high density planting is **Amarpali.**
149. Mango inflorescence contains, which type of flowers - **Male and Hermaphrodite.**
150. The off season mango variety is - **Niranjan.**
151. The only mutant cultivarr of mango is - **Rosica.**
152. Magnifera contains **about 41 species.**
153. The regular bearing varieties of mango are - **Neeum, Banglora, Pairi, etc.**
154. Mosambi a sweet orange cultivar was introduced from - **Mozambiue.**
155. The edible portion of citrus fruits is - **Juicy Placental Hairs.**
156. The first record of the name grape fruit in Jamaica was in - **1814.**
157. The Banana breeding was first stared at - **Imperial College of Tropical Agriculture in Trinidad (1922).**
158. Banana Breeding was stared in India in **1949 at Central Banana Research Station Aduthurai (T.N.).**
159. Banana can be stored at **- 13 C and 85% RH.**
160. Banana initiate flowering **9-12 months after Planting.**
161. The notice is papaya is **Tropical America.**
162. The yellow pigment in papaya is **Caricaxanthin.**
163. Papaya is the richest source of **Vitamin A** next to mango it contains 2020 IU/ 100 g fruit.
164. King of temperate fruit is - **Apple.**
165. Family of apple is **Rosaceae.**
166. Mother of all delicious group of cultivars is - **Red Delicious.**
167. The king of arid fruit is **Ber.**
168. The commercial propagation method off grape is - **Hard Wood Cutting.**
169. In India the black soil is mostly found in - **Maharashtra.**

170. Red soil is mostly found in - **Tamil Nadu**.
171. The soil which is most suitable for most crops is - **Sandy Loam**.
172. Red colour in the red soils is due to presence of various **Oxides of Irons**.
173. The vertical section of the soil showing the various layers from the surface to the unaffected parent material is known as - **Soil Profile**.
174. The arrangement of soil particles and their aggregate into certain defined is called - **Soil Structure**.
175. Particle density of soil is also known as - **True Density**.
176. Generally the particles density of normal soil is - **2.65 gram/cm^3**.
177. The bulk density of normal soil is - **1.33 g/cm^3**.
178. The formula of porosity of soil is - porosity = **100 (Bulk Density / Particle Density)**.
179. The cation exchange capacity (CEC) of montmorillonite is - **60-100 me/ 100 gm**.
180. The first soil classification system was developed by - **Russion Scientist Dokuchaev (1880)**.
181. The U.S classification system the soil orders are - **10**.
182. Formula of bulk density is = **wt. of soil / volume of solids and pores**.
183. Law of minimum was given by - **Justus Von Liebig (1840)**.
184. Top most mineral horizon is - **A horizon**.
185. Exchangeable sodium percentage **is also known as soluble percentage**.
186. The term pH was introduced by **Sorensen (1909)**.
187. Energy exchange elements are **H&O**.
188. K & Na is determined by **Flame Photometer**.
189. Micro nutrient is also known as - **Trace Element, Oligo-Element or Spurned Elements**.
190. Crop logging techniqueue was given by - **H.F. Clements**.
191. Metal nutrients are - **K, Fe, Ca, Mg, Mn, Zn, Cu**.
192. Highly mobile nutrients are - **N, P, K**.
193. Criteria of essentiality was proposed by **Arnon & Stout (1939)**.
194. The term fuctional nutrient was proposed by - **Nicholas (1961)**.
195. Moderate mobile nutrient is **Zn**.

196. FYM contains - **0.5% N, 0.2% P2O5 & 0.5% K_2O.**
197. The P is available at **6-7 pH**.
198. Bio fertilizers are - **Rhizobium, Azotobacter, Azospirillum, PSB, VAM, BGA, etc.**
199. Approx. 20 gram of rhizobium culture is required to treat **one kg of seed**.
200. One packet of rhizobium culture contains **20 gram of rhizobium culture**.
201. Azotobacter is used for - **Rice, Cotton, Sugarcane**.
202. Azospirillum biofertilizer used **for Sorghum**.
203. The cytoplasm and nucleus combinedly called - **Protoplasm**.
204. Cell theory was proposed by **M.J. Schleden and Theodo Schwann**.
205. The term Lysosomes was first used by **Dave in 1955**.
206. The main function of Golgi bodies are - **packaging of food materials such as protein, lipids etc**.
207. The function of ribosomes are - **protein synthesis**.
208. The term mitochondria was given by - **C Benda (1897)**.
209. The simple microscope was invented by - **Galileo (1610)**.
210. The shortest phase of all mitosis phase is - **Anaphase**.
211. Meiotic division is also known as **reducetional division**.
212. The term mitosis was coined by **Fleming (1882)**.
213. Mitosis occurs in **Somatic cells**.
214. Segregation occurs during **Meiosis**.
215. The longest mitotic phase is - **Prophase**.
216. Absorption Spectrum of photosynthesis **is blue and red**.
217. Photosynthesis is occur in present of Sunlight + Carbshydrate.
218. Dark reaction was discovered by **Blackman**.
219. Dark reaction is also known as - **Blackman reaction or C_3 – cycle**.
220. The main site for dark reaction of photosynthesis is – **stroma**.
221. The evaporation of water from plants is called - **transpiration**.
222. Evaporation from soil and transpiration combined are called **Evapotranspiration**.
223. C_4 Plants proposed by **Hatch and Slack (1970)**.

224. The dark reaction process of photosynthesis has been named variously - **Calvin Cycle or Blackman Reaction**.

225. Light reaction is also known as **Hill Reaction**.

226. In photosynthesis, the solar energy converts into - **Chemical Energy**.

227. The OSMOSIS term was given by - **Abble Nollet (1748)**.

228. The term "diffusion pressure deficit" was introduced by **- Meyer in 1938**.

229. The term Imbibition was coined by - **Sachs**.

230. Imbibition Pressure is also known as - **Matric Potential**.

231. Root pressure word was coined by- **Stephan Hales**.

232. Father of Physiology - **Stephan Hales**.

233. Soil less cultivation of plants is known as **Hydroponics**.

234. Rate of transpiration is determined by - **Potometer**.

235. The loss of sap (water) from the injured parts of plants is called - **Bleeding**.

236. Approximately 90% of the world photosynthesis is carried out by - **Marine and Fresh Water Algae**.

237. Net production of ATP in **Glycolysis is 2**.

238. Growth rate is measured by **Auxanometer & Cresco Graph**.

239. Photoperiodism term was coined by - **TD Lysenko (1920)**.

240. Apical Bud dominance is caused by which hormone - **Auxin**.

241. Which hormones is used as a herbicide = **Auxin**.

242. Which hormone is related to drought tolerance- **Absicic Acid (ABA)**.

243. Dormancy break by which hormone - **Cytokinis**.

244. Stress hardening in plants can be active by which hormone **ABA**.

245. Fruit ripening hormone is **- Ethylene**.

246. The term auxin was used by **- Frits went (Holland) 1905**.

247. Auxin is now known as - **Indole-3-acetic acid (IAA)**.

248. Which hormones are the Auxin **- IAA, IBA, NAA, 2-4-D, 2, 4, 5-T, MCPA, IPA TIBA etc**.

249. Shedding of plant parts is due to - **ABA hormones**.

250. Cycocel (CCC) is a **Growth Retardant**.

251. The most widely used auxin herbiceds are **2-4-D, 2, 4, 5-T, MCPA**.

252. Gibberellins were first discovered in **Japan**.

253. The hormones also known as - **Photo hormones, growth hormones, growth substances**.
254. Main auxin is also known as - **IAA**.
255. Root promoting hormones is **IBA**.
256. The precursor of IAA is - **Tryptophasne**.
257. Synthesis of IAA in plants requires **Zinc**.
258. The term phytohormone for hormones of plants is suggested by - **Thimann (1948)**.
259. The term genetic resources was coined by - **Frankel**.
260. Mendel began his famous experiments on **Pea in 1857**.
261. Mendel Paper entitled **"Experiments in Plant Hybridization" was presented in German Language in two papers in 1865 (Feb 8 and March 8)**.
262. Mendel Paper published in the annual proceeding of society in **1866**.
263. The term "Genetics" was coined by - **Bateson (1905)**.
264. The term gene, genotype and phenotype were used by - **Johannsen in 1903**.
265. Germplasm theory was proposed by **Weismann**.
266. Crossing over term was used first time by - **Morgan and Cttell in 1911**.
267. Operon model of gene regulation was discovered by - **Jacob and Monad 1961**.
268. The term chromosome was coined by - **Waldeyer in 1888**.
269. One gene one enzmes hypothesis was proposed by - **Beadle and Tatum in 1941**.
270. First time gene was divided into cistron, muton and recon by **Benzer (1955)**.
271. Sex chromosomes are also known as **Allosomes**.
272. Crossing over occurs in - **Pachytene stage**.
273. The tendency of two or more genes to remain together in the same chromsome during inheritance is referred to as - **Linkage**.
274. Chiasma type theory of crossing over was proposed by - **Janssens**.
275. Genic balance system of sex determination was proposed by - **Bridge in 1922**.

276. Monogenic sex determination is found in - **Asparagus, Papaya, Maize, Spinach etc**.

277. The homozygous progeny of a self - pollinated homozygous plant is known as - **Pure lines**.

278. The concept of purelines theory was developed by - **Johannsen (1903), Danish Biologist**.

279. Breeding method for self -pollinated crops are - **Pure Line Selection, Mass Selection, Progeny Selection, Bulk Method, Pedigree Method, Single Seed Descent Method, Back Cross Method**.

280. A multiline variety is a mixture of - **Several Pure Lines**.

281. Bulk method was first used by - **Nelson Ehle in 1908**.

282. The modification of bulk method is - **Single Seed Descent**.

283. In pure line Theory, Johannsen was working on - **Princess Variety of Rajma**.

284. Mass Pedigree Method was proposed by - **Harrington in 1937**.

285. The oldest method of selection is - **Mass Selection**.

286. Progeny test was developed by - **Louis de Vilmorin**.

287. Progeny Test is also known as - **Vilmorin Isolation Principle**.

288. Single Seed Descent Method applied for the first time in oat in 1965 by - **Graphius**.

289. The term recurrent selection was coined by - **Hull in 1945**.

290. The use of synthetic varieties for commercial cultivation was first suggested in maize by - **Hayes and Garber, 1919**.

291. Synthetic varieties are maintained by - **Open Pollination**.

292. A synthetic variety can be developed from-**Inbreeds, clones and open pollinated varieties**.

293. The general or basic concept in the development of synthetic varieties is - **Exploitation of hybrid vigour**.

294. The concept of disruptive mating and selection was developed by - **Mather (1953) and Thody (1958)**.

295. The Concept of dialled selective mating system was originally develped by - **Jonsen in 1970**.

296. The hetrosis is also known as - **Hybrid Vigour**.

297. The term heterosis was first used by - **Shull in 1914**.
298. Dominance theory was proposed by - **Devenport in 1908**.
299. The cross between inbred line and open pollinated variety was done in - **Top Cross Method**.
300. The resistance of host to the particular race of a pathogen is known as - **Vertical Resistance**.
301. Vertical resistance is governed by - **One or few genes**.
302. The resistance of host to the all race of a pathogen is known as - **Horizontal Resistance**.
303. The concept of gene for gene hypothesis was first developed by - **Flor in 1946**.
304. The disease resistance is governed by - **Several Genes**.
305. Who developed the concept of vertical and horizontal resistance - **J.E. Vender Plank (1963)**.
306. Father of genetic engineering is - **Pual Berg**.
307. Polymerase Chain Reaction (PCR) was developed by - **Dr. Karl Mullis (1980)**.
308. Golden rice was developed by - **Ingo Protrykus and Peter Bayer**.
309. The plant part, which is used for regeneration is called - **Explant**.
310. A mass of regenerated cells culture medium is called - **Callus**.
311. The suspension of free cells of callus in a liquid medium is known as - **Suspension Culture**.
312. The regeneration capacity or ability of a plant cell to develop into whole plant is known as - **Totipotency**.
313. What is the embryogenesis - It is the process of formation of somatic embryos from **Callus**.
314. The process of dirrerentiation of shoot & root from the somatic embyos is called - **Organogenesis**.
315. Regeneration of whole plant from anther is called - **Anther Culture**.
316. Haploid are developed by - **Anther Culture**.
317. The Blue biotechnology is a term that has been used to describe - **The marine and aquatic application describe**.
318. The biotechnology applied to agriculture process is known as - **Industrial Biotechnology**.

319. The first genetically modified food product was a - **Tomato, which was transformed to delay its ripening**.
320. M-RNA is primarily used for - **Transcription of DNA**.
321. RFLP stands for - **Restricting Fragment Length Polymorphism**.
322. Plant tissue culture was begun by - **G. Haberlandt (Geman Botanist) in 1808**.
323. research in the area of tissue culture was first started in india at - **Department of Botany, DU in 1960s**.
324. National Bureau of Plant Genetics Resources is situated at - **New Delhi**.
325. The project on Indian mustard oil with higher b-carotene is initiated by - **TERI**.
326. A genotype developed by the process of genetic engineering is called - **Transgenic**.
327. What is gene cloning - **A technique of genetic engineering, which is used to make several identical copies of a gene is called gene cloning**.
328. Which soil borne bacterium is used for developed of transgenic plants - **agrobacterium tumifaciens**.
329. In potato, the protein quality has been improved by - **transferring serum albumin gene from human**.
330. In tobacco, cold resistance power has been achieved by tranering gene from - **Arabidopsis thaliana**.
331. Little leaf of citrus is caused by - **MLO, S**.
332. Pahala blighted Sugarcane is caused due to deficiency of - **Mn**.
333. Reclamation disease is caused due to the deficiency of - **Cu**.
334. Late blight of potato is caused by - ***Phytophthora infestence*** **(fungi)**.
335. Bunchy top of banana is caused by - **Virus**.
336. Grassy shoot disease of s.cane is caused by - **Mycoplasma**.
337. Kresek in rice is caused by - ***Xanthomonas Oryzae***.
338. Heart rot of sugar beat is casued by - **Deficiency of Boron**.
339. Most dangerous disease of potato - **Late Blight of Potato**.
340. Philology disease in plants is caused by - **Mycoplasma**.
341. Club root of cauliflower is caused by - ***Plasmodiosphora brassicae***.

342. Sheath blight of rice is caused by - ***Rhizoctonia solani.***
343. Vector of leaf-curl disease in cotton is - **White flies**.
344. Orobanche is a parasite associated with - **Tobacco**.
345. Akiochi disease is due to - **Sulphur Toxicity.**
346. White rust of crucifers is caused by - ***Albugo candida*****, (fungi)**.
347. Ufra disease in rice is caused by- **Nematode**.
348. Whiptail of cauliflower is due to the deficiency of - **Mo**.
349. Vector of rice tungro virus is - **Green Leaf Hopper**.
350. Kresek symptom is found in - **Bacterial Blight**.
351. Loose smut disease of wheat is a - **Internally Seed Borne**.
352. Mantek Disease of rice is caused by - **Rice Root Nematode**.
353. Yellow leaves of tea occur due to the deficiency of - **Sulphur**.
354. Die back of shoots occur due to deficiency of - **Copper**.
355. **Die back** is the major disease of rose in India.
356. Crop showing maximum resistance to nematode is - **Marigold**.
357. The late blight disease destroy the potato crop of Ireland in **1845**.
358. Father of pathology is - **Anton de Berry**.
359. Father of mycology is - **P.A. Micheli**.
360. Life cycle of wheat rust is given by - **K.C. Mehta**.
361. Top sickness of tobacco is caused by - **Deficiency of Boron**.
362. The pathogen associated with the discovery of bordeause mixture is - ***Plasmosphora viticola***.
363. Loose smut of wheat can be controlled by seed treatment of - **Vitavax**.
364. Late blight of potato was introduced in India for first-time in-**Nilgiri-Hills**.
365. Black heart of potato is caused by - **Poor ventilation in storage (Lack of oxygen)**.
366. The tundu disease of wheat is usually associated with nematode known as - ***Anguina Tririci***.
367. Udabatta disease of rice is due to - ***Ephelis oryzae***.
368. Ring rust of Aonla is caused by - ***Ravenellia emblica*** **var.** ***fructoidae***.
369. Flag smut of wheat is caused by - ***Urouystis gramines***.

370. Leaf curl disease of tobacco is caused by virus and it is transmitted by - **White fly**.

371. Leaf blight of cotton is caused by - ***Alternaria macrospora***.

372. Father of Nematology - **Bastian**.

373. Father of Microbiology - **Louis Pasteur**.

374. Bacterium was discovered by - **Antony Van Leewenhock**.

375. fungicide used in the control of smut is - **Vitavax**.

376. Father of Modern Plant Pathology is - **Anton de Berry**.

377. Bordeaux Mixture was developed by - **P.M.A. Millardet**.

378. Common nematicides are - **DD Mixture, Phorate, Etylene Bromide, etc**.

379. Central Plant Protection Training Institute is situated at - **Hyderabad (Telangana State)**.

380. First plant protection adviser to govt. of India **- H.S. Pruthvi**.

381. Integrated Pest Control term was coined by - **Bartlett (1956)**.

382. The Term " Pest Management" was given by - **Geier (1970)**.

383. The blood colour of insects is - **Green with yellow**.

384. Honeybees belongs to the order of - **Hymenoptera**.

385. Entomology word was taken from - **Greek language**.

386. Father of Insect Physiology is - **Wiggelworth**.

387. The first pesticide was discovered by - **Millardet (1882)**.

388. DDT was discovered by - **Zedler scientist**.

389. First insecticide discovered was - **DDT**.

390. Plant Protection and Quarantine Act was pased in the year - **1912**.

391. Insecticide Act was passed in **1968**.

392. Destructive Insect and Pest Act was passed in - **1914**.

393. Maximum percentage of insecticide is used in which crop - **Cotton (54%).**

394. **"Warfarin"** is a rodenticide.

395. Contact insecticides are - **DDT, Malathian, Aldicarb, Monocrotophos, Aldrin, etc**.

396. Biopesticides are - **NPV, GV, *Bacillus thuringiensis* etc**.

397. Pest population should be kept below the **- Economic Threshold Level**.

398. Very safest insecticidede for honeybee - **Endosulphon**.

399. Rat poison is known as - **Zinc phosphide (Zn_3P_2)**.

400. Lead Arsenate was first used in **1832 by Moulton**.

401. Overhead cost is known as - **Fixed Cost**.

402. Price theory is a branch of - **Micro- Economices**.

403. Average cost is equal to - **Total cost / Output**.

404. Land rent is an example of **Fixed Cost**.

405. Prime cost is also known as - **Variable Cost**.

406. Opportunity coast is also known as - **Alternative Cost**.

407. The father of Economics is - **Adam Smith**.

408. Diminishing return is also known as - **Law of Variable**.

409. The most effective way to overcome the defects of agricultural marketing is - **Regulating marketing**.

410. The law of variable proportion is generally referred to as the - **Law of Diminishing Return Production is a Function of - Factors**.

411. AGMARK Act was enforced in year - **1937**.

412. **In mixed farming,** the contribution of livestock to gross farm income is at least.

413. Input -Output relationship is also known as **- Production Function**.

414. Those product, in which two product produced together are called **- Joint Product**.

415. Dumping activity is seen under **- Monopoly**.

416. Export - Import Bank of India was set up on - **1st January, 1982**.

417. IRDP was stared in the year of - **2 Oct, 1980**.

418. When marginal production is zero, then total production is - **Maximum**.

419. KVK is also known as - **Farm Science Centre**.

420. KVK was stared in the year - **1974**.

421. Operational Research Project (DRP) was stared in the year - **1974-75**.

422. Lab to land programme was started in the year - **1979 June 1**.

423. National Agri. Technology Project was stared in the year - **1998-99**.

424. KVK was recommended by - **Mohan Singh Mehta Committee (1974)**.

425. Community Development Project was stared on - **1952**.
426. National Extension Service (NES) was stared in the year of **1953**.
427. Firka Development Scheme was launched under the guidance of National Commission of Agriculture was set up by government of India - **1970**.
428. HYVP was started in the year of **1966**.
429. Marthendam programme was started by - **Dr. Spencer Hatch in 1921**.
430. Etawah Pilot Project was started by - **Mr. Albert Mayer in Sep 1948**.
431. Indian Village Service was started in 1945 by - **Mr. Arther T. Mosher and Shri BN Gupta**.
432. The aim of Etqwah Pilot Project is **- Introducing intensive work on the Rural Reconstruction Front**.
433. Intensive Agriculture District programme was popularly known as - **Package Programme**.
434. IRDP was lunched on - **2 Oct 1980**.
435. National Agricultural Research Project was launched in - **Jan, 1979**.
436. T&V system was evolved in - **1973 by Israeli Expert Daniel Benor At Turkey**.
437. Training of Rural Youth for Self Employment (TRYSEM) in **1979**.
438. Total number of KVK operating in the country till **Dec 2011-600**.
439. Antuadya Yojana was started in **2 Oct 1977**.
440. Sampoorna Grammeen Rojgar Yojna (SGRY) was started in the year of **2001**.
441. FCI was started in **1970**.
442. FISCAL commission was appointed in **1949**.
443. Multiple Cropping Programming was Started in **1966**.
444. First irrigation commission was appointed in **1901**.
445. First deptt. of agriculture was established **in 1881**.
446. The extension work must be based on the **- Need and Interest of People.**
447. Father of Extension - **J Paul Leagons**.
448. Father of Sociology - **August Commte**.
449. National Agricultural Innovation Project was stared in - **2005**.
450. Total number of national bureaux are - **Six**.
451. Formula of coefficient of variance is - **SD/Mean x 100**.
452. T test is generally used, **when sample size is < 30**.

453. The correlation coefficient lies between - **-1 to + 1**.
454. The experimental design, which simultaneously control the variation in two directions is known as **LSD**.
455. The basic principles of field experimentation - **1 Replication 2 Randomized 3 Local Control**.
456. The technique of analysis of variance was developed by - **R.A. Fisher**.
457. If fertility gradient of land is in two directions, the experimental design will be - **Latin Square Design**.
458. The most commonly used measure of control tendency is - **Arithmetic Mean**.
459. Cumulative Frequency is also known as - **Ogive**.
460. The term 'Kurtosis' was given by - **Karl Pearon (1906)**.
461. Laplace used the normal distribution in - **Analysis of Errors of Experiments**.
462. The important method of least squares was introduced by - **Legendre in 1805**.
463. The name "Normal distribution" was coined independently by - **Peirce Galton and Lexix, (1875)**.
464. Normal distribution is also known as - **Law of Error, Law of Facility of Error, Lapace's Second Law, Gaussion Law etc**.
465. Generalize mean is also known as - **Power mean or Holder Mean**.
466. The standard deviation is widely used measure of - **Variability or Disperson**.
467. The term standard deviation was first used by - **Karl Pearson in 1894**.
468. If mean, medium and mode is equal, then skewness will be - **Zero**.
469. Correlation's are useful because they can indicate a - **Predictive Relationship**.
470. Which ANOVA is used when the experimenter want to study the effects of two or more treatment variables - **Factorial ANOVA**.
471. Which ANOVA is used when the subjects are subjects to repeated measures, in which the same subjects are used for each treatment - **Two-Way Anova**.
472. The completely randomized design is considered to be most useful in situation where,
    1. The experimental units are homogeneous

2. The experimental are small such as Laboratory experiments
3. Some experimental units are likely to be destroyed

473. In field experiments the commonly used design is - **RBD**.
474. For chi- square test, the minimum sample size should be - **50**.
475. Define biometry - **It deals with observation with living thing**.
476. Mean , median and mode are equal in - **normal distribution**.
477. Most frequent number in a data set, is known as - **mode**.
478. The well-known example of skewed distribution is - **Personal wealth**.
479. Harmonic mean is the - **Reciprocal of Arithmetic Mean**.
480. The term z-test is often used to refer specifically to the - **One Sample Location Test**.
481. T Test is a - **Parametric Test**.
482. The goodness of fit test is also known as - **Kolmogorov-Smirnov test**.
483. The number of independent pieces of information that go into the estimate of a parameter is called - **Degree of Freedom**.
484. The typical symbol for degree of freedom is - **d.f.**
485. F-test was coined by - **George W. Snedecor, in honour of Sir Ronald A Fisher**.
486. The term "water harvesting" was first used by - **Myors**.
487. The force working with the attraction of **water molecules toward** each other is known as - **Cohesion**.
488. The force working with the attraction of **water molecules toward solid particles is known** as - **Adhesion**.
489. The water held tightly to the surface of soil particles by adsorption forces is known as - **Hygroscopic Water**.
490. Water held by force of surface as continuous film around soil particles and in capillary space is known as - **capillary water**.
491. Capillary water is held between tension of **31 and 1/3 atmosphere**.
492. The movement of water from the surface into the soil is Called - **Infiltration**.
493. The water moves freely in response to gravitational force and drain out of the soil is called as **Gravitational Water**.
494. The water between field capacity and permanent wilting point is called as - **Available water**.

495. Drip irrigation concept was proposed by - **Simcha Blass**.
496. One ha. Cm of irrigation is equal to **100000 litters**.
497. Major source of irrigation in India is **wells (52%)**.
498. Which irrigation method is widely used in fruit orchard - **Bain Method**.
499. The automatic method of surface irrigation is - **Coblegation**.
500. Leading state in drip irrigation in India is - **Maharashtra**.
501. Soil moisture is measured by - **Tension meter**.
502. Water requirement of irrigation wetland rice is about -**1500 mm**.
503. Total water requirement of sugarcane crop is **200-300 cm**.
504. In saline soil, the method for irrigation used is - **flood Method**.
505. PF value was first introduced by- **Schofield**.
506. Vertical movement of water in soil is known as - **Percolation**.
507. ET are measured by - **Lysimeter**.
508. Parsall flume is used for measurement of - **Water Flow**.
509. Sericulture is also known as - **Silk Farming**.
510. The rearing of silkworms for the production of raw silk is known as - **Sericulture**.
511. How much silkworms are required to produce **1 kg of silk 5500**.
512. How much people are engaged in various sericulture activities in the country - **60 lakh**.
513. Central Sericulture Research & Training Institute is situated at - **Mysore, Karnataka**.
514. Sericulture has been practiced for at least **5000 years in China**.
515. Apiculture is also known as **Bee Keeping**.
516. Apiculture word was taken from - **Latin world**.
517. A location where bees are kept is **called An Apiary**.
518. A beekeeper is also known as - **Apiarist**.
519. A beekeeper bee in order to - **Collect honey, bee wax, to pollinate crops, to produce bees for sale**.
520. What is beekeeping - **It is maintenance of honey bee colonies, commonly in hives by humans**.

521. The two economically important products of bee keeping is **Honey and Wax**.

522. Average yield of Indian bee is **- 6-8 kg / colony / year**.

523. The main freshwater fish are - **Cop and Catfish**.

524. The main brackish - water fish are **- Hilsa and Mullet**.

525. Central Institute of Marine Fisheries Research was established - **Kochi, Kerla**.

526. The Central Inland Fisheries Institute was established at - **Barrackpure (W.B.)**.

527. The Central Institute of Fisheries Technology was established at - **Wellindon Island near Kochi, Kerala**.

528. Hydroponics is also known as **- Soilless Culture**.

529. The Method of growing plants using minerals nutrients solutions, in water, without soil is called - **Hydroponics**.

530. In 1699 John Woodward published his water culture experiment with the crop of - **Spearmint**.

531. Hydroponics traditionally known as - **Solution Culture**.

532. First time who promoted that solution culture can be used for agriculture crop production - **Prof. William Frederich Gericks, University of California at Berkeley in 1929**.

533. The term Aquaculture for hydroponics was used for the first time **by Prof. William F.G.**

534. The Mutant Varity of rice is - **Jagannath**.

535. The Pecursor of vitamin - **A is - B-carotene**.

536. Azolla is a **- aquatic fern**.

537. The Blue revolution in India was started during - **Fifth Five Year Plan (1970)**.

538. Evergreen revolution is related to - **Sustainable Crop Production**.

539. Precision Farming is also known as - **Site Specific Management**.

540. SRI technique Was developed in - **Madagascar**.

541. The term Synecology was given by - **Schroter & Kirchner (1902)**.

542. Pusa Mukta is the cultivar of **Cabbage**.

543. World's super rice variety, which is suitable for saline / Alkaline soil conditions is - **Lunishree**.

544. NRCMAP is currently established at - **Anand (Gujarat)**.
545. IPR stands for - **Intellectual Property Right**.
546. Beaufort Scale is used to measure - **WIND pressure / strength**.
547. What is dioeciously? **Those plants which contain one male and female flowers**.
548. The element that play an important role in male and female protoplasm constituent is - **calcium**.
549. Wheat is a **C3 plant**.
550. The plant parasite nematode was first time reported by - **Kanche**.
551. Mean, Median and Mode is equal in - **Normal Distribution**.
552. The inflorescence of jowar is called - **Panicle**.
553. Gromer, SCU and CDU are the - **Slow Release Fertilizers**.
554. Low of tolerance was introduced by - **Shelford**.
555. End product of glycolysis is - **Pyruvate**.
556. define lithophytes - **The plant grown is rock cervices are called lithophytes**.
557. During glycolysis, the Net gain of ATP is - **2**.
558. Phosphorus is extracted by - **Olsen method**.
559. The correlation between price and demand of a commodity is - **Negative**.
560. The limit of partial correlation coefficient is - **-1 to +1**.
561. For reclamation of Saline and Alkaline soils used are - **Gypsum and Pyrite**.
562. The term ecosystem was coined by - **A.G. Tansley (1935)**.
563. What is parthenocarpy - **The development of fruit without fertilization**.
564. The development of embryo without fertilization is called - **Apomixes**.
565. Relative humidity is measured by - **Pshychrometer**.
566. Father of Indian palebotany is - **Birbal Sahni**.
567. The weed used for ornamental purpose is - ***Lantana Camaera***.
568. The core component of chlorophyll is - **Mg**.
569. The soybean is sown at the depth of - **3 cm**.
570. The most prominent soil of India is - **Alluvial soil**

571. Chaisma is seen during - Diplotene sub - **Phase**.

572. The current method used for control of bollworm in cotton is - **Bt. transgenic cotton produced**.

573. Photo-Respiration occurs in - **Chloroplast**.

574. Water use efficiency is highest in which plants - **CAM Plants**.

575. Chromosomal theory of inheritance was proposed by- **Sutton & Boveri**.

576. The price fixed by govt. is called - **Minimum Support Price**.

577. The present farming system of India has become - **Market oriented**.

578. Which chemical is mostly used for breaking tuber dormancy in potato- **Thiourea**.

579. The term water harvesting was first time used by - **Mayers**.

580. The first commercial hybrid cotton of the world is - **Hybrid-4**.

581. International Day for Biological Diversity - **22 May**.

582. Measuring of heights one by - **Clinometer**.

583. The Science City of India is **Bangalore**.

584. Bangalore is also known as **Garden city**.

585. Origin of potato is from - **Peru (S. America**).

586. Dormancy breaking hormone is - **Cytokinin**.

587. The element involved in energy transfer and storage in plants is - **Phosphorus**.

588. Fruit ripening hormone is - **Ethylene**.

589. Which is the antidote of insect poisoning - **Atropine**.

590. Minimum support price is formulated by - **CACP (total 28 crops)**.

591. Luxury consumption or maximum uptake nutrient is - **K**.

592. Crop canopy temperature is measured by - **Infra-red Thermometer**.

593. Secondary Plant nutrients are- **Ca, Mg, & S**.

594. Hulling percent in rice is - **65%**.

595. Which method of sugar cane sowing prevents the lodging - **Trench method**.

596. Inflorescence of sugarcane is called - **Arrow 600.** Shedding of plants part is due to - **ABA hormone**.

601. First hybrid in the world of red gram is - **ICPH-8**.

602. The most preferable soil structure for agriculture is - **Granular & Crumby**.
603. Taxonomically organic soils are - **Histosols**.
604. Which soil is most suitable for dry land agriculture - **Black soil**.
605. First man made cereal is - **Triticale**.
606. Dead heart and white ear disease is related to - **Rice**.
607. IBPGR is situated at - **Italy, Rome**.
608. India's share in the fruit production in the world is **10%**.
609. What is relay cropping - **The second crop sown before harvesting of first crop**.
610. Measurement of sunlight intensity is expressed in - **Lux units(old)/ Einstein (new)**.
611. In water logged soils gas found in abundant - **Methane**.
612. Curing process is related to the crop - **Tobacco**.
613. The book "Nature & Properties of Soil" is written by - **N.C. Brady**.
614. Chemical used for delinting of cottons - $\mathbf{H_2SO_4}$.
615. Maximum N loss in rice crop is due to - **Denitrifiaction**.
616. Wind velocity can be measured by - **Anemometer**.
617. Queen of spices is - **Cardamom**.
618. Buck wheat is called - **Psedo cereal**.
619. Very dangerous disease of **sugar cane is red rot**.
620. World's staple food is - **Rice**.
621. Fibre of cotton contains - **Cellulose**.
622. Movement from higher concentration to lower concentrationn is known as - **diffusion**
623. Gold of America is known as - **Soybean**.
624. Formula of Net Capital Ratio is - **Total Assets /Total Liabilities**.
625. Theory of evolution was given by - **Charles Darwin**.
626. Which Indian Scientist shared World Food Prize for miracle maize - **Dr. Surinder K. Vasal**.
627. ICAR day is celebrated on **16 July**.
628. Light red carnation represent to - **Admiration**.
629. ANOVA table stands for analysis of - **Variance**.

630. Formula of Harvest Index is - **Economical yield/ Biological yield multiply by 100**.
631. RBI was established in **1935**.
632. Who discovered vitamin - **Funk**.
633. Regur soil refers to - **Black soil**.
634. Biological active form of glucose is - **D-form**.
635. Lodging preventor is - **CCC (Cycocel)**.
636. FAO was started in **1948**.
637. Which design is used when fertility gradient is in two direction - **LSD**.
638. International pest is ***Schistacerca gregaria***.
639. Bhoodan movement was started in **1951**.
640. Cooperative movement was started by- **F. Nicholus (1904)**.
641. Flared square symptoms are seen in cotton due to the - **Spotted Boll Worm**.
642. CAN is a **Neutral fertilizer**.
643. Root promoting hormone is **IBA**.
644. Ooze test is done for **detecting Bacteria**.
645. First product of urea hydrolysis is - **Ammonium Carbamate**.
646. NABARD was set up on the recommendation of - **Siva Raman Committee, Narsimhan Committee 1982**.
647. The trade name of BHC is - **Grammexone**.
648. Who discovered Mitocondria - **C. Benda (1897)**.
649. Bt cotton was discovered in - **1901 Japan by Ishwata**.
650. The term cytokinin was proposed by - **Letham (1963)**.
651. Kinetin and Zeatin are the - **Cytokonins**.
652. Photosynthesis is an - **Oxidation - Reduction Process**.
653. Largest importer of cut flower in the world is - **Germany**.
654. Gibberellin was discovered for the first time used by - **Kurosowa**.
655. Water soluble vitamin is - **Vitamin-C**.
656. Greenhouse effect is also known is - **Global Warming**.
657. VAM act as a - **Symbiotic Association**.
658. LEISA stand for - **Less External Input Sustainable Agriculture**.

659. Blue Revolution in India was started during - **Fifth Five Year Plan (1970)**.
660. Who discovered pH concept - **Sorenson**.
661. DRIS concept given by - **Beaufils**.
662. The term 'remote sensing' was first used in - **USA**.
663. Acidity in gram leaves is due to - **Malic Acid / Oxalic Acid**.
664. Photosynthetic rate is highest in **C4 plant**.
665. Lock and Key Model was proposed by - **Fisher**.
666. Anti - Sterility vitamin is - **Vitamin E**.
667. Double seed formation in cotton is due to - **Pink Bollworm**.
668. Forest Conservation Act was made in - **1980**.
669. Netural parthenocarpy is found in - **Banana**.
670. The term vernalization was coined by - **TD Lysenco**.
671. Chemical relationship is done between two cells by **Plasmodesmata**.
672. White revolution was accomplished by which scientist - **V. Kurien**.
673. How much time is needed for sterilization in autoclave - **15 minutes**.
674. In which bacterial reproduction phase, virus is involved - **Transduction**.
675. First plant flower mandi was situated at - **Bangalore.**
676. Regional Plant Resource Centre is situated at - **Bhubaneswar**.
677. Cultivation of crops in areas where the annual rainfall is less than 750 mm is called - **Dry Farming**.
678. Central Soil Salinity Research Institute is situated at - **Karnal**.
679. What is the ideal location of greenhouse - **North-South**.
680. According to ICMR vegetable required per capita per day is - **300 gram**
681. Biggest tree of the work is - **Banyan tree**.
682. Kisan call cente begun in the year **2004**.
683. Net cultivated area of India is - **143**.
684. Kisan divas are celebrated on **23 Dec**.
685. The share of agriculture in global warming is - **Approx. 25%**.
686. Environment day is celebrated on **- 5 th June**.
687. Which element is responsible for nodulation in legumes - **Phosphorous**.
688. Fingerprinting is related to which crop - **Pineapple**.

689. Yellow vein mosaic virus disease is related to **Bhindi**.

690. Which clay Mineral is dominant in Black soil - **Montmorillonite**.

691. Sprouting in onion is controlled by the hormone of - **MH**.

692. Which chemical/ hormone is used for regular bearing in mango - **Paclo butrozol (PP333)**.

693. High yielding variety of wheat was developed by - **Norman Borlaug.**

694. Break down of clods , crusts and plant material by the impact of particles moved by wind in saltation is called as - **Abrasion**.

695. A soil with a pH value less than 7 hence having more hydrogen ions than hydroxyl ions in the soil solution is called as - **Acid Soil**.

696. The movement of ions and water into the plant root because of metabolic processes by the root, frequently against an electrochemical potential gradient is referred to as - **Active Absorption**.

697. The chemical in commercial product that is directly responsible for the herbicidal activity is called its - **Active Ingredient**.

698. The transfer of heat energy by means of horizontal mass motions through a medium is termed as - **Advection**.

699. A technique in growing plants where in the plants derive their nutrients and eate from a mist of air and aqueous solution that comes in contact with the roots is called - **Aerophonics**.

700. A term proposed by Mc Gillchrist in 1965, which is a measure of how much a relative yield increase in component a is greater than for component b is known as - **Aggressively**.

701. The removal of excess water known as free water or gravitational water from the surface of the farm land as to create favourable soil conditions for plants growth is called as - **Agriculture drainage**.

702. The growing of agricultural crops along with the forest crop is known as - **Agri-silviculture**.

703. A land units defined in terms of major climate and growing period, which is climatically the homogenous response of a crop or a farming system is called as - **Agro-Climatic Zone**.

704. An alternative renewable fuel, produced from vegetable oils or animal fats through a refinery recess called trans- esterification is called as - **Bio-Diesel**.

705. In India, ICAR- Directorate of Medicinal & Aromatic Plants Research is located at - **Anand**.

706. A branch of science, which deals with the stud of grasses, their classification, management and utilization is called - **Agrostology**.

707. The area that comprises of all the areas from which water flows out into a common river or water pool is called as - **Catchment Area**.

708. A branch of science, that studies the aspects of climate, which are relevant to the problems of agriculture is called - **Agro-Climatology**.

709. A Medium, which resists change or a substance added to resist change in concentration of-r pH activity in a solution as called as - **Buffer**.

710. The crop which is grown as a substitute for the main crop that has failed on account of unfavourable condition is called as - **Catch Crop**.

711. The ratio of total cropped area in different seasons to the total land area is known as - **Cropping Intensity**.

712. The branch of agriculture which deals with scientific crop production and management is - **Agronomy**.

713. Cultivation of crops in Areas where rainfall is less than 750 MM per annum is known as - **Dry farming**.

714. The study of the inter - relationship of living organisms with each other and with their environment in an agricultural system is called as - **Agro-ecology**.

715. The crops that are grown to break the continuity of the agro-ecological situation of the field under multiple cropping systems are called as - **Brake Crop**.

716. The competitive interaction between the crop species in intercropping and sequential cropping through the release of chemical substances or toxins is called - **Allelopathy**.

717. The chemical substance released by one species may inhibit more strongly plants of the producer species itself is termed as - **Auto-inhibition**.

718. The production technique that combines agriculture and forestry, horticulture or animal husbandry on the same piece of land in the order to fully utilize the natural resources is called as - **Agro-Forestry**.

719. The fraction of the total solar radiation incident on a body that is reflated by it and expressed in % age is called as - **Albedo**.

720. A substance added to a soil for the improvement of its physical and chemical properties is called as - **Ameliorant**.

721. An appropriate combination of farm enterprises i.e. cropping system,

livestock, poultry, fisheries, forestry and the means available to the farmer to raise them for increasing profitability is called as - **Farming Systems**.

722. The scientist, who has been called as father of organic farming is - **Albert Howard**.
723. The process of use of micro- organism to remove salts from soil is referred as - **Bio-remediation**.
724. The organization that is related to export of 'Agricultural processed product is - **Agri-Export Zone**'.
725. "Sugarcane Breeding Institute" is located at **- Coimbatore**.
726. The ratio between water deficit to potential evapotranspiration and expressed in percent is termed as - **Aridity Index**.
727. A term given by Donald in 1963, Which is a measure to find out the yield of various crops when grown together as well as separately and indicating the yield per plant of different crops in mixture and their respective pure stand on an units basis is known as - **Competition Index**.
728. A biological model which is expected to perform or behave in a predictable manner a defined environment is known as **- Ideotype**.
729. Gene responsible for dwarfing characters in rice is - **Dee-Gee-Woo-gene**.
730. Hue denotes **Dominant Spectrum**.
731. Indian Institute of Soi Scienceis established at - **Bhopal**.
732. Directorate of Rupeseed & Mutard Research is located at **Sewar, Bharatpur, Rajasthan**.
733. Potassium is also known as **quality element**.
734. International Potash Institute is situated at **Basel (Swizerland)**.
735. Dr. Rudolf Steiner conceptualized and advocated organic agriculture as a trademark Demeter in 1927
736. One straw revolution book was written by MASANOBU FUKOKA in 1975.

2

Agronomy

1. **Agronomy** is derived from **Greek word**.
2. Indigenous plough cuts '**V-shape furrow**'.
3. Indigenous plough is a **primary tillage implement**.
4. The **ploughing depth of chisel plough** is 60-70 cm.
5. **Hoe is** an example of **secondary tillage implement**.
6. **Tillage improves** soil aeration.
7. Jethro Till is the **father of tillage**.
8. **Summer ploughing controls** problematic *Cyperus rotundus* L.
9. The **correct time for ploughing** depends on soil moisture.
10. **Depth of ploughing** depends on the depth of effective root zone.
11. **Country plough is** a multipurpose plough.
12. The **concept of minimum tillage** is originated from USA.
13. **Secondary tillage operations** are performed in the row zone.
14. **Zero tillage** is the last form of minimum Tillage.
15. **Jethro Tull coined** the word 'Zero tillage'.
16. **Tilth is the qualitative characteristics** of crumby condition of the soil.
17. **Crop growing Monsoon season** is June to July.
18. **Crop growing winter season** is October to November.
19. **Crop growing summer season** is February-June.
20. **1 Degrees brix is equal to** 1 g sucrose in 100 g of solution.
21. **Brix is important** in sugarcane.
22. Sugarcane and sugar beet is **biennial crop**.
23. **Dry farming** is based on natural precipitation.
24. Sugarcane and cotton are examples **cash crops**.

25. Cowpea and sesame are examples of **catch crops**.
26. Sunhemp is an example of **nurse crop** in sugarcane field.
27. Maize and greengrams are examples of **companion crops**.
28. Zinzvo is an example of **pasture crop**.
29. Zethro Tull defines **weed** as an **undesired plant**.
30. The weed ***Argemone mexicana*** causes harm to human health.
31. **Crop rotation** effectively controls *Avena fatua* and *Cuscuta spp.*
32. In **stale seedbed** crops germinate in weed free environment.
33. **AC Case inhibitors affect** cell membrane production in grass.
34. **ALS inhibitors affect** grasses and dicots alike.
35. **Glyphosate** is an **EPSPS inhibitor**.
36. Roundup is the **brand name of glyphosate**.
37. **Glyphosate is produced by** Monsanto.
38. 2, 4 – D is an **example of synthetic auxin herbicides**.
39. Triazine is an **example of photosystem II inhibitor**.
40. Paraquat is an **example of photosystem I inhibitor**.
41. **Atrazine** is commonly used in sorghum and corn.
42. Pendimethalin is a **pre-emergence herbicide**.
43. Fluazifop is a **post emergence herbicide**.
44. Paraquat and glyphosate are **non-selective herbicides**.
45. The **criteria of essentially** was given by Arnon, 1954.
46. **Average concentration of N in plant tissue** is 1.5%.
47. **Average concentration of P and Mg in plant tissue** is 0.2%.
48. **Average concentration of N in plant tissue** is 1%.
49. **Average concentration of B in plant tissue** is 20 mg/kg.
50. **Average concentration of Mo in plant tissue** is 0.1 mg/kg.
51. N, P and K are **highly mobile elements in plant**.
52. Ca and B are **immobile elements in plant**.
53. Urea is an **example of straight fertilizer**.
54. **Urea is also known as** carbamide.
55. **First laboratory synthesis of urea by**: Friedrich Wohler.

56. **CAN have neutral reaction effect** on soil.
57. By nature **potassium chloride is neutral**.
58. EDTA is a **synthetic chelating agent**.
59. DAP contains 18% N.
60. Borax contains 10.5% boron.
61. **Only N containing fertilizers** are urea and ammonium chloride.
62. Wheat and mustard are **example of companion crops**.
63. Safflower is an **example of guard crop** for pea.
64. Japanese mustard is an **augmenting crop** for berseem.
65. Cotton – berseem is an **example of relay cropping**.
66. Urd with maize is an **example of parallel cropping**.
67. Mustard with sugarcane is an **example of Multi-storeyed cropping**.
68. Sugarcane and potato are an **example of synergistic cropping**.
69. Alley crops are also known as **hedge row intercrops**.
70. Rabi crops are **long day plant**.
71. **Test weight** is the weight of 1000 seeds
72. Botanically, **cereals are caryopsis**.
73. **2n of rice** is 24.
74. Panicle is the **inflorescence of rice**.
75. The **SRI method** was developed in 1983 in Madagascar by Henri de Laulanie.
76. **Plant to plant distance in SRI** is 25 cm.
77. **Ruralia commoda** is the manual by Pietro de Crescenzi.
78. **Hulling percentage of rice** is 70-75%.
79. IR-8 is **miracle rice**.
80. **Aroma in rice** is due to Di-acetyl 1 propaline.
81. **Khaira disease in rice** is caused by the deficiency of Zinc.
82. Maize is known as the **Queen of Cereals**.
83. Maize is a C4 plant.
84. *Zeamayssacchrata* is **sweet corn**.
85. **Seed date of hybrid maize** is 20-25 kg/ha.

86. **Nicotine content in *Nicotiana rustica*** is 3.5 to 8%.
87. **Fire curing** is done for bidi.
88. Dough is the **harvesting stage of oat**.
89. Time for **adsali sugarcane** is July-August.
90. **Retting** is an important process in jute and linseed.
91. Lint percent in seed cotton is 33.
92. **Regur soil** is best for the **cultivation of cotton**.
93. Safflower is grown for oil.
94. **Protein content of soybean** is 40-42%.
95. Groundnut is a modified fruit.
96. Botanical name of blackgram is *Phasiolus mungo*.
97. Pigeonpea has the **lowest harvesting index** among pulses.
98. **Kidney disorder** is related with barley.
99. Gluten is the wheat protein.
100. **Shelling percentage of wheat** is 60%.
101. **Contingency irrigation** is life saving irrigation.
102. Flooding is the oldest form of irrigation.
103. Drip irrigation is originated form Israel.
104. There is no irrigation required at field capacity.
105. **Drip irrigation** method has the highest irrigation efficiency.
106. The maximum water requirement of sugarcane is similar to rice.
107. Annual rainfall, **dry farming**: <750 mm.
108. Cycocel and kaoline are **antitranspirents**.
109. Increased production is the goal of **intensification**.
110. Sustainable agriculture is based on ecosystem.
111. 'Agriculture' word is derived from **Latin word (agri+culture)**.
112. A branch of agricultural science which deals with principles and practices of soil, water, and crop management is **Agronomy**.
113. "Agronomy" word is derived from word. **Greek (agros+nomos)**.
114. The non-edible oilseed crops are.**Castor and Linseed**.

115. The Indian originated field crops are **Arhar, Mung, Urd, Cotton, Jute, Kodo, Kutki, Oat etc**.

116. Kharif crops are generally denoted as Short **day plants (NOTE: Short-day plants are also called as Long-night Plants. Long-day plants are also called as Short-night Plants. Short-day plants usually flower in the early spring or autumn**).

117. Rabi crops are generally denoted as. **Long day plants (NOTE: Long-day plants flower when they are exposed to day length Longer than the critical minimum period**).

118. **Day-neutral plants are Cotton, Maize, Sunflower, Safflower, Groundnut, Buckwheat, Tomato**.

119. The optimum time of sowing for Kharif crop is **June-July**.

120. The optimum time for Rabi crop **Last week of October to the first week of November**.

121. The optimum depth of sowing for most of the field crops **3-5 cm**.

122. The recommended fertilizer dose (N:P:K) for cereal crops are **4:2:1**.

123. The C_4 Plants are **Maize, Sorghum, Bajra, Sugarcane, Millets**.

124. The recommended fertilizer dose (N:P:K) for pulse crops are **1:2:1 or 1:2:2**.

125. The recommended fertilizer dose (N:P:K) for oilseed crops are **3:2:1**.

126. The recommended fertilizer dose (N:P:K) for fodder and fibre crops are **2:1:4**.

127. The C_3 Plants are Rice, **Wheat, Barley, Pea, Gram, Mustard and Rye, Cotton, Arhar, Soybean, Sunflower, Lentil, Sugarbeet, Tomato etc**.

128. The CAM Plants are **Pineapple, Khajur, Cactus, Sisal**.

129. The optimum temperature for better crop production is between **18-240**.

130. The weight of 1000 seeds of a crop? **Test weight**.

131. The weight of 100 seeds of a crop? **Seed Index**.

132. The net assimilation rate is express in terms of **g cm^2 day^{-1}**.

133. Which of the following crop geometry ensures uniform solar radiation availability to crop? **Square**.

134. Wavelength longer than______ m/μ is not visible to the eye, and are called infrared. **750**.

135. The development stage of a plant after which no further increase in dry matter occurs in the economic part is known. **Physiological maturity**.
136. Crops that are cultivated to catch the forthcoming season when main crop is failed? **Catch/Contingent Crops**.
137. A close-growing crop, grown primarily to improve and protect the soil from erosion through their ground-covering foliage and/or rootmats between periods of regular crop production? **Cover Crops**.
138. When both main and intercrop is benefited to each other? **Complementary Crops**.
139. The crops leave the field exhaustive after growing? **Exhaustive Crops**.
140. Any crop or combination of crops is grown for grazing or harvesting for immediate or future feeding to livestock? **Ley Crops**.
141. Such crops are grown to conserve the soil moisture through their ground-covering foliage? **Mulch Crops**.
142. The seed of succeeding crops is sown broadcast at 10 to 15 days before harvesting rice crop? **Paira/Utera Crops**.
143. Generally, the third row of the crop is removed or growing of crop in pair row and the third row is escaped with an object to conserve the soil moisture in Dryland areas? **Paired row Crops**.
144. Such cops are neither complementary nor competitive? **Supplementary Crops**.
145. Crops, those are grown to protect the main cash crop from a certain pest or several pests? **Trap Crops**.
146. Cereals are botanically **Caryopsis**.
147. The more nutrient exhaustive family is **Poaceae (Graminae)**.
148. The non-conventional oilseed crop is **Sunflower**.
149. Two major commercial crops are (i) **Cotton** (ii) **Sugarcane**.
150. When both - main and intercrop is benefited to each other? **Complementary Crops**.
151. Who is the father of tillage? **Jethro Tull**.
152. The primary tillage implements are **Deshi plough, MB plough, Ridge plough, Disk plough etc.**.
153. The mechanical manipulation of the soil is k/s as **Tillage**.
154. The good physical condition of soil after tillage is **Tilth**.

155. The tillage operation mainly aims to break, open and turn the soil **Primary/ tillage**.

156. The tillage referred as no tillage in which the crop is planted in unprepared soil **Zero tillage**.

157. The word 'Zero tillage' was termed by **Jethro Tull**.

158. A system of tillage in which organic residues are not inverted into the soil and used as a protective cover against erosion and evaporation losses of soil moisture? **Conservation tillage or stubble mulch tillage**.

159. Conservation tillage tends to encourage **Higher microbial population**.

160. The tillage implement used to break subsoil is **Chisel plough**.

161. Ridge plough is used for **Earthing-up and form ridges and furrows**.

162. Star weeder is used for **Weeding in dry lands and groundnut fields**.

163. Disc plough is used for **Deep ploughing in grassed field**.

164. The secondary tillage implements are **Cultivator, Harrows, Hoe, Planker, Roller etc**.

165. The optimum range of available soil moisture for convenient and effective plowing **50-75%**.

166. An ideal condition of soil for crop growth? **Seed-bed**.

167. Conventional tillage involves **Minimum tillage, Zero tillage, and Conservation tillage**.

168. The tillage aims to reduce tillage to the minimum necessary for ensuring a good seed-bed, rapid germination, satisfactory stands and favourable growing condition? **Minimum tillage**.

169. Harrows are used for **Preparation of seedbed, destroy weeds**.

170. Mechanization index is found highest in **Wheat crop**.

171. The tillage operation, done to create a good seedbed for proper seeding/ planting **Secondary tillage**.

172. The botanical name and family of paddy Oryza sativa, Poaceae.

173. The botanical name and family of wheat ***Triticum spp., Poaceae***.

174. The botanical name of maize is **Zea mays**.

175. Botanical name of sorghum is ***Sorghum bicolor***.

176. The botanical name of two-rowed barley and Six rowed barley is ***Hardium distichoum*** and ***Hardium vulgare, respectively***.

177. Botanical name of Kabuli/White Chickpea? ***Cicer kabulium***.

178. Botanical name of Desi/Brown Chickpea? ***Cicer aeritinum***.
179. Early maturing pigeonpea/Arhar is **Cajanus cajan flavus**.
180. Latest botanical name of mung/Greengram is ***Phasiolus aureu***.
181. Botanical name of Fieldpea ***Pisum sativum var. arvense***.
182. Botanical name of Garden pea ***Pisum sativum var. hartens***.
183. Latest botanical name of Urd/Blackgram is ***Phasiolus mungo***.
184. The botanical name of soybean is ***Glycine max***.
185. The botanical name of sunflower ***Helianthus annus*** .
186. The botanical name of Brown/Indian mustard ***Brassica juncea***.
187. The botanical name of sarson is ***Brassica compestris***.
188. Bunch/Spanish/Erect type groundnut is ***Arachis hypogea fastigat***.
189. Spreading/Verginia runner type groundnut is ***Arachis hypogea procumbens***.
190. The botanical name and family of Flex/linseed is ***Linum usitatisium, Linaceae***.
191. Indian/old world cottons are ***Gossipium arborium, G. herbacium***.
192. American/new world cotton is ***G. hirsutum***.
193. Egyptian cotton/sea island cotton is ***G. barbadence***.
194. The botanical name of safflower ***Carthamus tinctorius***.
195. The botanical name of white jute ***Corchorus capsularis***.
196. The botanical name of tropical cane ***Saccharum officinarum***.
197. The botanical name of berseem/Egyptian clover is ***Trifolium alexandrinum***.
198. The botanical name of lucerne/Alfalfa is ***Medicago sativa***.
199. The botanical name of oat crop is ***Avena sativa***.
200. The botanical name of potato is ***Solanum tuberosum***.
201. The origin place of potato is **South America (Peru)**.
202. The botanical name of tobacco is ***Nicotiana tabacum***.
203. How many agro climatic zones are identified by the Planning Commission (1985-90)?**15**.
204. Indian Institute of Wheat and Barely is located at? **Karnal**.
205. Protein content in Rice is? **6-7**.

206. "SRI" stands for?- **System of Rice Intensification**.

207. Who is called Father of Agronomy? - **Peter Decresenzi**.

208. Founder of Relay cropping is? **Dr. Biwant Raj**.

209. The response of plants to the relative length of day and night is knownas? **Photoperiodism**.

210. Which of the following crops protect soil surface from erosion? **Cover crops**.

211. Who among the following divided the world in 6 Biogeographical regions **Alfred Russel**.

212. The Planning Commission (1985-90) divided the country in how many agro climatic zones? **-15**.

213. Formation of soil is known as - **Pedogenesis**.

214. The smallest 3D volume of a soil needed to give full representation of horizontal variability of soil is termed as- **Pedon**.

215. Total amount of water present in the soil is called? **Holard**.

216. Deficiency of which element in soil causes chlorosis in plants? - **Magnesium**.

217. The formula for balanced salt solution of mineral elements for plant growth was given by? **Knop**.

218. Which of the following are critical elements? **-N, P, K**.

219. Which element is common to nucleic acids, phospholipids, ATP and ADP? **P (Phosphate)**.

220. Which group is required for chlorophyll synthesis? - **Fe, Mg**.

221. Which element is a part of cytochrome oxidase? - **Fe**.

222. Which of the following contains highest nutrients? - **Droppings of sheep & goat**.

223. What is the drawback of poultry manure? **It ferments very quickly**.

224. Trench method of FYM preparation is also known as - **Dr. C. N. Acharya method**.

225. Which of the following is suitable method for application of FYM? **To spread uniformly on the surface of soil**.

226. Which of the following is not a straight fertilizer? - **FYM**.

227. NPK Composition of standard fertilizer mixture for Potato and Paddy **is - -6,12,6**.

228. Which commercial fertilizer contains Mg & Ca as secondary nutrients? **Epsum & Gypsum**.

229. Application of manures and fertilizers to the established crop within crop duration is termed as - **Top dressing**.

230. Thousands of hectare of cultivable land has been abandoned due to infestation of ***Cyprus rotundus***.

231. Which of the followings weed is indicator of poor light soil? - **Cymbopogon**.

232. Heavy infestation of which of the following causes poor ploughing performance - ***Cynodon dactylon***.

233. Which of the following is not a indigenous weed? **- *P. hysterophorus***.

234. Total root parasite on tobacco is - **O. Cermua**.

235. Eichhornia crassipes is - **Floating weed**.

236. Zygogramma bicolorata beetle feed on leaves of **Parthenium**.

237. Phytophthora palmivora causes root rot in - **Morrenia Odorata**.

238. Application of volatile chemical on confined spaces to produce gas that will destroy weed seeds is called**? Fumigation**.

239. A herbicide which kills some species in a mixed growth plant species are termed as - **Selective herbicides**.

240. Which of the following is long period residual? - **Atrazine**.

241. "Paraquat spray" can be inactivated by spraying - **1% ferric chloride**.

242. Compounds which aid the herbicide availability at the action site in plants are called? **Adjuvant**.

243. Which of the following thickening agents are used with herbicides? - **Sodium alginate**.

244. When the resistant plant posses two or more distinct resistance mechanism to a **Multiple resistance**.

245. Which of the common crop rotation under irrigated fields condition of Coimbatore district? - **Ragi-Cotton-Sorghum**.

246. Foliage applied herbicide in - **Glyphosphate**.

247. Water content of apical portion of roots and shoots is - >**90%**.

248. Majority of rainfall in India is contributed by **South west monsoon**.

249. The methods adopted to measure the evapotranspiration is **Lysimeter experiment**.

250. Check basin method of irrigation is also known as - **Beds and channel method**.

251. Which method of irrigation is suitable for orchard crops? - **Ring basin method**.

252. Average rate of application of water to crop is called as - **Precipitation intensity**.

253. Water spread area of a sprinkler is given by equation- (where, R = Radius of wetted area (md = diameter of nozzle, h = pressure head at nozzle) - **R = 1.35 x Underoot of (dh)**.

254. In flaccid cell, the pressure potential is - **Zero**.

255. Which of the following helps plant in reducing epidermal transpiration? - **Cutin and Cuticle Both**.

256. Which bacterium is responsible for denitrification in N2 cycle? - ***Bacillus subtilis***.

257. To explain ascent of sap in plants, Sir J.C. Bose (1923) proposed - **Vital theory**.

258. What is the effect of atmospheric pressure on transpiration**? It increase slowly**.

259. Maximum density of water is at - **-4°C**.

260. Water absorption through roots increases **when - Transpiration increases**.

261. South western monsoon crops are also knows as - **Kharif**.

262. **Sugarbeet** is the frost tolerant crop?

263. **Bajra and Jowar both** is drought tolerant crop?

264. **Sugarcane and Sunflower both** long duration crop (above 150 days)?

265. How many principles are there for plant distribution ? **-8**.

266. Injury caused by high temperature on the sides of bark is known as - **Sun sclad**.

267. For which of the following,very high relative humidity is harmful? - **Sunflower and Tobacco both**.

268. Grain filling at high humidity results in - **Reduced crop yield**.

269. Which of the following is not a Heliophyte? **Turmeric**.

270. The response of plants to the relative length of day and night is known as - **Photoperiodism**.

271. In ________ erosion resistance crops and erosion permitting crops are grown in alternate strips. - **Strip cropping**.

272. Which of the following is anti-transpirant? **Both Phenyl mercuric acetate and Alkanyl succinic**.

273. When trees are introduced into farms along with field crops, it is known as - **Both Agrisilviculture and Agro forestry system**.

274. The growth rate is highest in - **Red light**.

275. Which of the following is a natural growth regulator? - **Ethylene**.

276. Synthesis of amylase in certain cereal grain is stimulated by - **GA**.

277. Only gaseous phytohormone is - **Ethylene**.

278. Tuberization in potato can be achieved by **ABA**.

279. Sugarcane bowl of world: **Cuba**.

280. Water requirement of sugarcane: **2200 mm**.

281. Sugarcane is a **Heavy feeder crop**.

282. Zero tillage is practiced in **sugarcane**.

283. Indian cane: ***saccharum sinensis* (Chinese cane), *saccharum barberi* (Indian cane)**.

284. Tropical cane: ***saccharum officinarum*** (noble cane).

285. Wild cane: ***saccharum robustum***.

286. Indian Institute of Sugarcane Research, **Lucknow, 1952**.

287. Sugarcane Breeding Institute, **Coimbatore, Tamil Nadu, 1912**.

288. Sugarcane Breeding Institute, **Regional Centre Karnal, 1932**.

289. The minimum soil temperature for the germination of sugarcane sets is in the range of **19-21°C**.

290. Ripening temp: **12-14°C**.

291. Optimum temp: **28-32°C**.

292. Formative stage: **26.5°C**.

293. In India, Sugarcane is grown as a Kharif Crop. It needs hot and humid climate with an average temperature of **21°C to 27°C**.

294. Setts: 3 budded **35000-40000 setts/ha**.

295. 2 budded **80000 setts/ha**.

296. 1 budded **120000 setts/ha**.

297. Spacing for cane- **75-90 x 45 cm**.
298. Autum: **Oct-Nov (north india, 12 months)**.
299. Eksali: **Jan-Feb (north india, 12 months)**.
300. Adsali: **July-Aug (south india, 18 months)**.
301. Flatbed method: **in north India, sowing depth 8-10 cm**.
302. Ridge and Furrow method: **in south India, depth 10-15 cm**.
303. Trench method: **prevent to lodging**.
304. Partha method: **in south India**.
305. Spaced method: **used single budded setts**.
306. Winter nursery method: **used 3 budded setts**.
307. Rayangan method**: used for gap filling**..
308. Skip furrow method: **combine method of flat bed and trench method**.
309. Pit method: used in sloppy field with pit size **20-30 cm**.
310. Germination stage: **0-60 DAP**.
311. Formative stage: **60-130 DAP (most critical stage)**.
312. Grand phase: **130-250 DAP**.
313. Maturity phase: **250-365 DAP**.
314. Tying: **bind stems together**.
315. Earthing up: **plant root covered by soil to prevent logging. Done in 45 DAP**.
316. Logging: **nutrient status of plant**.
317. Treshing: **removing dry leaves from crop**.
318. Blind hoeing: **weeding after sowing and before germination**.
319. Arrowing: **inflorescence of cane is arrow and process of emerging is called arrowing**.
320. Water lodging: **crop fall down due to excessive water**.
321. Crop lodging: **crop fall down due to high wind velocity**.
322. Logging term given by **H.F. clament**.
323. Bagases**: used for fuel and paper making**.
324. Molasses: **for Alcohol production**.
325. The yield of molasses is approximately **3.0 percent per ton of cane**.

326. Press mud: **reclamation of acidic soil. (Due to lime contain)**.
327. Gasohol: **petrol (80%) + alcohol (20%)**.
328. Origin of pigeon pea: **Africa**.
329. Botanical name of pigeon pea: ***Cajanus cajan***.
330. Family of pigeon pea: ***leguminaceae / papilionaceae***.
331. Chromosome number of pigeon pea: **2n=22**.
332. Germination type of pigeon pea: **hypogeal**.
333. Protein in pigeon pea: **21-25%**.
334. Inflorescence of pigeon pea: **axillary raceme**.
335. Other name of pigeon pea: **red gram, Congo pea, no eye pea, Arhar, tur**.
336. Deep rooted system in **pigeon pea crop**.
337. Pigeon pea is an **Often-cross pollinated crop**.
338. Pigeon pea is a **C_3, short day plant**.
339. 1 gram of protein = **1 gram of carbohydrate**.
340. Stems are woody**, branchy can go up >4**.
341. Pigeon pea is a **Hardy crop, most drought tolerant crop**.
342. Harvesting index of pigeon pea, **lowest among cereals 19%**.
343. Storage moisture percentage of pigeon pea: **10-11%**.
344. Test weight of pigeon pea: **72g**.
345. Fruit type of pigeon pea: **pod**.
346. Edible part of pigeon pea: **seed**.
347. Water requirement of pigeon pea: **300-350 mm**.
348. Cajanus cajan var. flavus (tur) is a - **Short duration**.
349. Pods are light coloured with **3-4 seeds**.
350. Annuals, yellow flowers, **fewer plain pods**.
351. Plant height **2-3 meter**.
352. Early maturity **(100-150 days)**.
353. No red **vines present**.
354. Agronomy One-liner Flowering in **Dec-Feb**.
355. Pod maturity at different time Cultivated in **Southern India**.

356. *Cajanus cajan* var. bicolour (arhar) is a **Long duration, flowers yellow with purple streak**.
357. Std petals bears red veins on **dorsal side**.
358. Pods are dark coloured with **5-7 seeds**.
359. Cultivated in **northern India**.
360. Flowering **Sep-Nov**.
361. Pod maturity at same **time (syncronization)**.
362. Plant hight **5-6 meter**.
363. Late maturity **(150-180 days)**.
364. It is also found suitable for forage, as **cover crop, shade and hedge**.
365. Temp. Required to pigeon pea at Germination: **30-35°C**.
366. Vegetative growth: **20-25°C**.
367. Flowering and pod development: **14-18°C**.
368. At maturity: **35-40°C**.
369. Origin of black gram**: India (central Asia)**.
370. Botanical name of black gram: ***Vigna mungo***.
371. Family of black gram**: *Leguminosae***.
372. Chromosome number of black grams: **2n=22**.
373. Inflorescence of black gram: **Auxiliary racemes**.
374. Another name of black gram**: urd bean**.
375. Germination type of black gram: **epigeal**.
376. Test weight of black gram: **36-39g**.
377. Black gram is a Self-pollinated, **C3, short day plant**.
378. Black gram is a Mainly **tropical plant**.
379. Fruit type of black gram: **pod**.
380. Edible part of black gram: **seed**.
381. Water requirement to black gram: **250-300 mm**.
382. Protein% in black gram**: 24%**.
383. Seed multiplication ratio of black gram: **1:40**.
384. Seed rate of black gram: **20-25 kg/ha**.
385. Optimum temperature required for black gram: **27-30°C**.

386. Origin of green gram: **Central asia**.
387. Botanical name of green gram: ***Vigna radiate***.
388. Agronomy One-liner Green gram is Commonly called as **Mung bean**.
389. IIPR is located at - **Kanpur**.
390. Green gram plant **Deep rooted**.
391. Seed multiplication ratio of green gram: **1:40**.
392. Origin of chickpea**: India, southwest Asia**.
393. Botanical name of chickpea: ***cicer arietinum***.
394. Family of chickpea: ***Leguminosae/papilonaceae***.
395. Chickpea fruit is known as **Pod**.
396. Germination of chickpea: **hypogeal**.
397. Father of Agronomy is - **Peter Decresenzi**.
398. **Dr. Biwant Raj** is the founder of Relay cropping.
399. The Indian Institute of Farming System Research is located at - **Modipuram, Meerut, U.P.**
400. Scientific study of mixed cropping was firstly done by - **La-Flitze (1928)**.
401. Relay cropping concept was derived from - **Relay race**.
402. Natural Resource management is an example of **Sustainable Agriculture**.
403. Cropping System is a component of - **Farming System**.
404. When the sub crop are sown to supplement of yield of the main crop, the sub crops are known as - **Augmenting Crops**.
405. The main crop is grown in the center, surrounded by hardly or thorny crops, the sub crop is known as - **Guard Crops**.
406. The yield of both crops , grown together are found to be higher than the yield of their pure crops on unit area, such cropping is known as - **Synergetic Cropping**.
407. What is a **Cash Crop** - those crops, which may be sold directly from the field without processing like vegetables, potato, etc.
408. Formula of **Cropping Intensity** = Total Cropping Area / Total Sown Area X 100.
409. The growing of two or more crops on the same field in a year is called - **Multiple Cropping**.

410. The practice of raising grasses or pasture, is also known as - **Ley farming**.

411. In India, the **maximum acreage and production of wheat is in U.P.**

412. The type of fruit of wheat is **Caryopsis**.

413. The optimum temperature for **ideal germination of wheat is 20 to 25°C**.

414. Norin-10 gene was brought to USA by - **S.C. Salamone (1984)**.

415. The dwarf winter variety of wheat first time developed through using Norin gene by - **Dr. O.A. Vogel (USA)**.

416. Triple gene dwarf variety were released in 1970.

417. **Double gene dwarf varieties of wheat** - HD 2204, Janak, UP 215, Pratap, Kalyansona, Arjun, Sonara 64.

418. **Single gene dwarf varieties of wheat** - Sonalika, Girija, UP-262, Wl 711, Lerma Rojo 64A.

419. The protein found in wheat flour which is helpful for chapati making - **Gluten**.

420. Central Zig-Zag axis of wheat is called - **Rachis**.

421. Mexican dwarf wheat was evolved by - **Dr. N.E. Borlaug**.

422. The toxic alkaloid found in the leaves of sorghum crops is - **Dhurin or HCN**.

423. **CSH 1** was released as **first dwarf sorghum hybrid** in 1964.

424. Which crop is known as camel crops - **Sorghum**.

425. Native of Jowar is **Africa** (Ethiopia & Sudan).

426. The **maximum area & Production of Jowar is in Maharashtra**.

427. Jowar is poor in lysine but **rich in - Leucine**.

428. The inflorescence of Jowar is called - **Panicle**.

429. **First hybrid variety of Bajra is HB-1** (1965) developed at PAU, Ludhiana.

430. Protein content in Bajra is **11-12%**.

431. Origin of Bajra is **Africa**.

432. Bajra is sensitive to – **Water logging and Acidic soil**.

433. The **largest producer state of cotton in India is - Gujarat**.

434. Bad opening of cotton balls is known as - **Tirak**.

435. In cotton, the maturity of fibre is judged by - **Arealometer**.

436. H-777 is the cultivar of bt-**Cotton**.

437. **Hybrid-4** is the **first hybrid commercial cotton** of the world developed by - **Dr. C.T. Patel in 1970**.

438. The leaves of cotton becomes **reddish in colour due to decrease nutrient uptake and moisture**.

439. **DRIS** concept given by - **Beaufils**.

440. Cotton belongs to- **Malvaceae Family**.

441. **Intra-Specific hybrid** of cotton is - **Hybrid-6, Savita, Surya**.

442. **Interspecific hybrid** of cotton is - **Varalaxmi, DCH-32, HB-224 etc**.

443. One seed of cotton has **60000-80000 fibres**.

444. **Ginning %** = wt. of lint / wt. of seed cotton taken for ginning x 100.

445. The removal of lower leaves which come in contact with soil and lose their commercial value is known as **Priming Process**.

446. Priming method is used in - **Cigarette and Wrapper Tobacco**.

447. **The main aim of Topping and Desuckering** is to divert the energy and nutrient of plant from flower to leaves.

448. The removal of the lateral branches or suckers or auxiliary buds is called - **Desuckring**.

449. Central Tobacco Research Institute is situated at - **Raja-h-Mundri, A.P. (1947).**

450. The richest source of protein among the food grain is - **Pulses**.

451. Head quarter of Directorate of Pulses Research is located in - **Kanpur, U.P.**

452. Which pulse crop doesn't fix Nitrogen from atmosphere - **Rajma**.

453. The word Pulse is derived from - French world '***pottage***' or '***poride***' means - Soup (dal).

454. The term ***Legume*** is taken from a Latin word which means 'to collect'

455. India is the largest producer of - **Gram**.

456. The sour taste of gram leaf is due to presence of - **Maleic acid (60%)** and **Oxalic acid**.

457. Shelling % of Pea is **49%**.

458. Harvest index of Arahar is **19%**.

459. Origin of Arahar is from - **Africa**.

460. Queen of cereals is called **Maize**.

461. **The composite cultivars of maize are** - Kisan, jawahar, Amber, Vijay, Sona, Vikaram, etc.

462. Maize is a **C_4 Plant**.

463. In India the rank **first state in area and production of Maize is U.P.**

464. The maize Protein is called - **Zein**.

465. **Male** inflorescence of maize is called - **Tassel**.

466. The **critical stage for water application in maize** is - **Tasseling and Silking stage**.

467. Which Indian scientist shared **World Food Prize for Miracle Maize - Dr. Sruinder K. Vasal (2000)**.

468. Origin of Maize is from - **Mexico**.

469. All India Coordinated Maize Improvement Project was started in - **1957** at New Delhi.

470. Double cross technique for hybrid seed production was developed by - **D.F. Jones (1920)**.

471. The idea of hybrid maize was first conceived by: **E.M. East and G.H. Shull (1910)**.

472. The criteria for harvesting of sugarcane - **Withering of lower leaves, brix 20%, sucrose 15%**.

473. **Align planting method** of sugarcane is evolved by - **Sam Higginbottom University of Agriculture, Technology and Sciences (Deemed to be University) Prayagraj, Uttar Pradesh**

474. Adhsali sugarcane is common in **Maharashtra**.

475. **Sugarcane is planted in furrows** because - it reduce lodging.

476. Inflorescence of sugarcane is generally called - **Arrow or Open Panicle**.

477. Sucrose content in cane is - **13-14%**.

478. Sugar from juice is - **6-10%**.

479. **Hand Refractometer reading for sugarcane maturity is 20** (Brix Value).

480. Sugarcane Breeding Institute is situated at - **Coimbatore**.

481. Adhsali Sugar cane crop mature in 18 months.

482. Father of hybrid rice is - **Yuan Long Pin**.

483. India's first dwarf rice variety released during 1968 was **Jaya**.
484. World's first high yielding semi dwarf basmati rice variety is - **Pusa Basmati-1**.
485. World's first super rice variety for saline / alkaline soil - **Lunishree**.
486. Harvesting of paddy is done at moisture content - **21-23%**.
487. First mutant variety of rice is **Jagannath (Gamma ray irradiation of T-141)**.
488. Which gas is released from field of paddy - **Methane (CH4)**.
489. **SRI technique** was originated in Madagascar in 1987 by **Prof. Henry De Laulanie**.
490. First super fine aromatic basmati hybrid is - **Pusa RH-10**.
491. Oilseed production programme was started during - **1990-91**.
492. Technology mission on oilseed was established on **1986**.
493. The Bio-Herbicide used in rice crop is **Collego**.
494. Golden rice is rich in **Vit. A**.
495. Rice is having protein content of **6-7%**.
496. Most widely used post-emergent herbicide in rice **is Propanil**.
497. Indian Institute of Wheat and Barely (DWR) is located at **- Karnal**.
498. The sowing depth of dwarf wheat varieties is - **4-5 cm**.
499. Water requirement in Maize **500-800 mm**.
500. Average Maize yield in India is **2500 kg/ha**.
501. National Research Centre for Groundnut established at - Junagarh (Gujarat).
502. The Kernel to nut in shell in groundnut is 70 percent.
503. The oil to nut in shell in groundnut is 28 percent.
504. The NBPGR is located at New Delhi.
505. The Marsh spot of peas is due to Mn deficiency.
506. The phahla blight of sygarcane is due to Mn decicionely.
507. The SKUAST-K is situated at Shalimar, Srinagar.

3

Soil Science

1. The fine earth covering land surface acts as a reservoir of nutrients and water **Soil**.
2. The word 'Soil' is derived from **Latin**.
3. Study of origin, classification, morphology of soil is known as **Pedology**.
4. The study of soils in relation to crop growth **Edaphology**.
5. The father of Soil Science **Dokuchalev**.
6. The concentration of soil water in soil **50%**.
7. Organic matter content in Indian soil is **5 %**.
8. Natural soil aggregates/mass are known as **Peds**.
9. The science describes rocks **Petrology**.
10. Granite and Basalt are **Igneous rocks**.
11. Lime stone, Sand stone and Dolomite are **Sedimentary rocks**.
12. Gneiss, Marble, Quartzite and Slate are **Metamorphic rocks**.
13. The rocks gets broken in pieces due to temperature is called **Exfoliation**.
14. Feldspar, Quartz and Mica are **Primary minerals**.
15. Kaolinite, Halloysite and Dickite are **1:1 type silicate clay minerals**.
16. The example of 2 : 1 type silicate clay minerals are **Montmorillonite, Vermiculite and Illite**.
17. Chlorite is **2:1:1 or 2:2 type clay mineral**.
18. The most dominant mineral on earth crust **Feldspar (48%)**.
19. The weathering mineral, having most stable soil structure **Kaolinite**.
20. Which mineral is a source of phosphorus and boron in soils? **Apetite**.
21. The hydroxide act as cementing agent in binding the soil particles together **Fe and Al**.

22. A vertical section of soil through all its horizons is called **Soil Profile**.
23. Which horizon is called Fertile zone? **'A' horizon**.
24. The horizon absent in arable land **'O' horizon**.
25. The eluviation horizon is **'E or A2' horizon**.
26. The illuviation horizon is **'B' horizon**.
27. A+B horizons are collectively called as **Solum**.
28. A+B+C horizons together called as **Regolith**.
29. The formulae of bulk density of soil **BD (g/cc) =Wt. of oven dry soil/ Volume of soil(solid+pores)**.
30. Bulk density of normal soil is **1-1.6 g/cc**.
31. The formulae of particle density of soil PD (g/cc) =Wt. of oven dry **soil/ Volume of soil solid**.
32. Widely accepted fixed value of particle density Is **2.65 g/cc**.
33. The soil having PD 2.50 g/cc and BD 1.25 g/cc will have __________ % porosity. **50**.
34. A field soil sample weighing 60 g, lost12 g on over dying. What is the moisture percent on dry weight basis? **25%**.
35. The weight of one hectare of surface soil (O-15 cm) in kilograms 2.24 × 106 kg/ha.
36. The arrangement of primary particles of soil **Soil structure**.
37. The relative proportion of sand, silt and clay is termed as **Soil texture**.
38. The best agricultural soil structure is **Crumby/Spheroidal**.
39. Soil structure proving less porosity in soil **Platy**.
40. The best agricultural texture is **Loam**.
41. NBSS and LUP centre is located at **Nagpur**.
42. Commonly followed soil particle classified in India is **International Society of Soil Science (ISSS)**.
43. According to IISS, the particle size of course sand **2-0.2 mm**.
44. The particle size of fine sand **0.2-0.02 mm**.
45. The particle size of silt **0.02-0.002**.
46. The particle size of clay < **0.002**.
47. The maximum pore space are found in **Clay soil**.

48. Soil colour is determined by **Munsell Colour Chart**.
49. It is the relative purity or strength of the spectral colour **Chroma**.
50. The capacity of the soil to change its shape under moist conditions **Soil Plasticity**.
51. The attraction of solid surface for water molecules is called as **Adhesion**
52. The density of soil water is maximum at **4°C**.
53. The surface tension of water is at 25°C **72.7 dyne/cm^2**.
54. Solution whose strength or concentration is accurately known is termed as **Standard Solution**.
55. Water held between 1/3 rd and 15 atm is termed as **Available water**.
56. Father of Agricultural chemistry - **J.V. Liebig**.
57. "Central Soil Salinity Research Institute" situated at - **Karnal (Haryana)**.
58. Who is considered as father of soil testing - **Troug**.
59. Law of restitution was proposed by - **Liebig**.
60. Which horizon zone of Washing out - **A (Eluvial)**.
61. Which horizon zone of Washing in - **B (Illuvial)**.
62. Red soil mainly formed by which soil order - **Alfisols Soils**.
63. Black soil mainly formed by which soil order - **Vertisol soil**.
64. 'Largest Soil group is **- Alluvial soil group (143 mha)**.
65. C:N ratio of micro organism – **4:1 to 9:1**.
66. Equivalent acidity of Anhydrous ammonia - **148 meq/100 gm**.
67. Equivalent acidity of Urea – **80-85**.
68. Equivalent acidity of DAP – **77**.
69. "International potash of Institute" situated in – **Switzerland (1952)**.
70. Red colour of soil due to – **Hematite**.
71. Red yellow color of soil is associated with – **Ferric oxide**.
72. Which soil warmed up quickly – **Sandy soil**.
73. The B horizon is generally absent in which soil **Black soil**.
74. Soil fertility and fertilizers was written by – **Samuet L. Tisdale**.
75. Which micro-nutrient are non-metal element – **Boron**.
76. Intracellular deposits of which element is known as plant opals ? **Silicon**.

77. "Indian Society of Soil Science" (ISSS) established in which year - **Dec. 1934 (Kolkata)**.
78. "International Soil Reference and Information Center" situated in - **Netherland (1966)**.
79. "International Union of Soil Science" founded in - **1924 (Australia)**.
80. For nitrification, the optimum pH is - **8.5**.
81. White alkali soils are - **Saline Soil**.
82. Humus fraction lightest in colour is - **Fulvic acid**.
83. Pedoturbation is the process of - **Mixing of the soil**.
84. In India, which soil order have negligible area - **Histosols**.
85. Particle density of clay soil is - **2.8**.
86. "Indian Institute of Soil Science" is situated in - **Bhopal (MP)**.
87. "Indian Institute of soil and water conservation" is situated in - **Dehradun**.
88. First production of fertilizer in India - **Single Super phosphate (1906)**.
89. Calcareous soil is formed in arid as well as - **Humid region**.
90. Generally plant loses water through - **Transpiration**.
91. Active soil forming factor are - **Organism & Climate**.
92. Which group of microorganism have add highest Biomass in soil - **Fungi**.
93. Soil color determination by - **Munsell's Chart**.
94. Cation exchange is an important component of - **Soil fertility**.
95. Positive Bacteria color is - **Red**.
96. "Sulfur Institute" situated in - **Washington DC (1960)**.
97. First product of Urea if apply in soil - **Ammonium carbonate**.
98. Which is best method for application Fe and Zn - **Foliar Application**.
99. Which is best suitable method for application Ca, B, S, and Mo - **Soil Application**.
100. Kandic and Arfullic origin found in which order - **Alfisol**.
101. Surface color of Alkali soil is - **Black**.
102. Lowest category of soil classification includes - **Series**.
103. Sugarbeet is an indicator plant for - **Sodium**.
104. Wheat is an indicator plant for - **Copper**.
105. Chemical formula of siderite is - **$FeCO_3$**.

106. Hue denotes **Dominant spectrum.**
107. Phosphorus uptake in alkali soil in the form of **PO_4^{3-}.**
108. Fertilizer produce acidity in soil **Ammonium sulfate.**
109. Maximum N content found in N fertilizer **Anhydrous ammonia.**
110. Rock phosphate has P_2O_5 **20-40%.**
111. The conversion factor for calculating P from P_2O_5 **P_2O_5 x 0.44.**
112. The phenomenon slickenside found in —— soil **Vertisol.**
113. —— organic fraction not soluble in **both acid and alkali-Humin.**
114. In munsell colour chart hue 10 represents as ______ **Yellow.**
115. pF scale for Field capacity is _____ 2.5.
116. Available water held between **Field capacity to wilting point.**
117. The central cation on silica tetrahedron is _____ **Si^{4+}.**
118. ______ plant is used for indication of permanent wilting point - **Sunflower.**
119. Mica is a type of clay mineral **2:1.**
120. Hydrogen bond found in which clay mineral **Kaolinit.**
121. Base saturation in lateritic soil **> 40%.**
122. Immobilization of sulphur occur in soil when C:S ratio is _____ **> 400:1.**
123. Denitrification is a process of ______ **Reduction.**
124. Microorganism involve in conversion from nitrite to nitrate **Nitrobactor.**
125. Thiourea is type of fertilizer **Artificial nitrification inhibitors.**
126. Chemical formula of thiourea is _____**$SC(NH_2)_2$.**
127. N content in thiourea is _________ **36.8%.**
128. Quantity of soil to be taken for texture analysis **20g.**
129. The K_2O content in Illite varied between **6 to 10%.**
130. The microorganism lies below < 10°C is called __ **Psychrophiles.**
131. Thermophiles organism lies in temperature **> 45°C.**
132. Optimum temperature for mesophile **20-40°C.**
133. _________ Organism derive their energy from oxidation of complex organic compound **Heterotrophs.**
134. _________ Organism utilize carbon from CO_2 for their cellular synthesis **Autotrophs.**

135. Most of the soil organism is ________ **Mesophiles**.
136. Optimum growth of bacteria in soil **Neutral to Slightly Alkaline**.
137. Fungi growth optimum in pH **4.5-6.5**.
138. Actinomycetes prefer which condition for their growth **Slightly Alkaline**.
139. Aroma of earthy smell in first shower of monsoon is due to ________ **Actinomycetes**.
140. ——— genera of bacteria found maximum in soil - **Bacillus**.
141. Rhizobium is an _____ organism - **Heterotrophs aerobic**.
142. Nitrification process ceases at pH < **5**.
143. Denitrification mostly takes place in _____ **Anaerobic**.
144. Nitrification rapid in _____ season **Summer**.
145. The peak microbial population was observed in soil generally after addition of residue **1-2 week**.
146. For mineralization of P optimum ratio of C:N:P should be _____ **100:10:1**.
147. Ammonia volatilization is purely process of _____ **Chemical**.
148. —— element suppressed Downey mildew of pear millets - **Chloride**.
149. Improper development of wheat inflorescence is due to deficiency of _____ **Boron**.
150. Organic matter obtained **OC x 1.724**.
151. Saturated hydraulic conductivity is measured by _____ **Amoozemeter**.
152. Hydraulic conductivity is found to decrease in the following order **Ca - soil > K - soil > Na - soil**.
153. ________ chemical compound is used for improve soil aeration, infiltration and water holding capacity of soil **Jalshakti**.
154. Reddish colour in muriate of potash is due to _____ **Impurities**.
155. Feldspars is primary mineral that occurs pre-dominantly in ________ **Igneous rock**.
156. CO_2 content in rainwater is _______ 0.45%.
157. In which point resistant to wind erosion in soil **Permanent wilting point**.
158. Bedrock is absent in which soil? **Alluvial soil**.
159. Saline soil also known as _____ **Solonchalk**.
160. Kari is the local name of _____ **Acid sulfate soil**.

161. The honey-comb structure found in _____ **Laterite soil**.
162. _______ soil occur mainly a soil crust problem - **Silty clay loam**.
163. Red colour of soil is due to _____ **Hematite**.
164. Black colour of soil is due to _____ **Titaniferous Magnetite**.
165. Total soil order found in world - **12**.
166. Flow of nitrogen in soil mainly due to _____ **Mass flow**.
167. Micronutrient flow is mainly a _____ process in soil **Diffusion**.
168. When organic matter added into the soil effect on bulk density **Decreased**.
169. ________ factor is more responsible for black colour in kanhar (Vertisol) soil? **Titaniferous Magnetite**.
170. Which factor is responsible for red colour in bhata (Entisol) soil - **Hematite**.
171. Yellow colour appear due to hydration of hematite found in which soil? **Matasi (Inceptisol)**.
172. Soil order found in Chhattisgarh **5**.
173. Which Soil order is newly identified in Chhattisgarh? **Mollisol**.
174. Which soil has low water and nutrient holding capacity of Chhattisgarh? **Bhata**.
175. Bhata soil is also known as _____ **Gravely**.
176. Matasi soil is also known as _____ **Sandy loam**.
177. Dorsa soil is also known as ____ **Clay loam**.
178. Kanhar soil is also known as _____ **Clayey**.
179. ________ soil covers maximum area in Chhattisgarh? **Matasi**.
180. Bhata soil covers area in Chhattisgarh of total are **12%**.
181. Dorsa soil covers area in Chhattisgarh of total area **27%**.
182. Kanhar soil covers area in Chhattisgarh of total area **21%**.
183. Total soil order found in India **12**.
184. Which one is lower category for classification of soil - **Soil Series**.
185. ___________ clay mineral is responsible for cracking in black soil? **Montmorillonite**.
186. pH meter measures **H ions activity**.
187. EC (electrical conductivity) meter measures **Soluble salt in soil**.

188. The unit used for the measurement of electrical conductivity is ___**dS/m**.

189. Olsen method is used for phosphorus estimation **Neutral to alkali soil**.

190. Which method used for determination of available phosphorus in acid soil **Bray**.

191. $KMnO_4$ (potassium permanganate) is used for_____ **Available nitrogen**.

192. Which instrument is used for the estimation of potash (K) **Flame-photometer**.

193. Spectrophotometer is used for determination of____ **P**.

194. Reclamation disease due to copper deficiency occurs in plants grown on newly reclaimed **Aridisols**.

195. Peat soils are generally **Black, Heavy and Acidic**.

196. A branch of science that deals with the study of origin of the soil, its classification and description is known as **Pedology**.

197. Muck soil containing organic matter **20-50%**.

198. A soil having pH 6.0 is called **acidic soil**.

199. Solum is made up of **A horizon**.

200. CaCO3, (lime) is added to reclaim of acidic soil.

201. Alkali soil are reclaimed by the addition of **gypsum**.

202. Cation exchange capacity is highest of **vermiculite**.

203. The vertical section of soil, which shows different horizons, is called **soil profile**.

204. Geological erosion is also known as **natural erosion**.

205. _____ is the content of organic matter in alluvial soil? **0.34-0.77**.

206. 'Humus' is made up of **humic materials and non humus material only**.

207. A soil with 20 me/100g CEC and 5 me/100g exchangeable H will have a base saturation of **25%**.

208. Soils with high pH are generally deficient in **Zn and Mn**.

209. Chlorite belongs to the **2, 2 type**.

210. Soil layer of maximum leaching is indicated by the symbol of **A2**.

211. The mineral chlorite has an extra layer of **illite**.

212. ______ Minerals is the main source of boron in soils? **Tourmaline**.

213. Symbiotic association of fungal hyphae with plant roots is referred to as **Mycorrhiza**.

214. The capacity of soil to be molded when wet, i.e., to change shape in response to stress and to maintain that shape when the stress is removed is termed **Plasticity**.

215. Deforestation has an alarming **effect on soil erosion**.

216. The occurrence of quartz in high amounts makes a rock **acidic**.

217. The concept of pH is suggested by the scientist **Sorenson**.

218. In soil Taxonomy udolls is a suborder of **Mollisols**.

219. Silica-sesquioxide ratio is useful in the evaluation of soil in respect of **iron and aluminum content**.

220. In which horizon of soil minerals are found **B**.

221. ———— of the following type of soil will form a colloidal solution when suspended in water? **Clay**.

222. The proportion of sand, silt and clay in a soil sample determines **texture**.

223. Availability of water and minerals in the soil for maximum absorption by roots is in the **A-horizon**.

224. Soil conservation measures are mainly aimed at protecting _____ **Topsoil**.

225. Soil is formed by __________ of rocks **weathering**.

226. A dark soil is usually rich in __________ **organic matter/humus**.

227. The physiochemically active portion of soil is __________ **top soil**.

228. Soil erosion can be stopped by __________ **plantation**.

229. Erosion involves movement of _______ from one place to other. **Soil**.

230. Soil is of different types: clayey, loamy and __________ **sandy**.

231. _______ is a section through different layers of the soil. **Soil profile**.

232. ________ soil is used to make pots, toys and statues **Clayey**.

233. The mixture of rock particles and humus is called **soil**.

234. ———— has smallest size soil particles? **Clay**.

235. Cotton grows well in **black soil**.

236. ———— cannot prevent soil erosion - **Deforestation**.

237. The topmost layer of soil is known as **top soil**.

238. Sandy soils are found in **deserts**.

239. Less humus and more minerals occur in the **C-horizon**.
240. _______ has the smallest size of particles - **Clay**.
241. ______ type of soil is the best for growing cotton? **Sandy-loam**.
242. ________ do not cause weathering - **Clouds**.
243. ________ horizon consists of small lumps of rocks with cracks and crevices. **C-horizon**.
244. Water logging can be expected in the soil which is rich in ______ **clay**.
245. Paddy grows best in _______ **clayey soil**.
246. ________ The breaking of rocks by the action of wind, water and climate is called **weathering**.
247. The layer which is soft, porous and can retain more water is **topsoil**.
248. B-horizon is also known as _________ **sub-soil**.
249. _______ is removed during soil erosion. **Top soil**.
250. The soil with large amount of fine particles is called _________ soil. **Clayey**.
251. The soil with a mixture of humus, sand and clay is called ________ soil. **Loamy**.
252. _______ soil is fit for cultivation. **Loamy**.
253. Percolate means __________ or __________ gradually **filter, ooze**.
254. Soil __________ water and also holds it. **Absorbs**.
255. The soil with larger particles in it is called ________ soil. **sandy**.
256. _________ horizon of soil contains very less organic matter. **B**.
257. Soil is formed by _________ of rocks.**weathering**.
258. The upper crust of earth, which is capable of supporting vegetation is called ________ **soil**.
259. A vertical section through different layers of the soil is called the _________ **soil profile**.
260. _________ soil has least water holding capacity. **Sandy**.
261. Accumulation of wastes in the soil generated by human activity which alter the features of soil is called _________ **soil pollution**.
262. The dark colour of the top soil is due to the presence of _______ in it. **Humus**.

263. The uppermost layer of the earth's crust which is composed of a thin layer of mineral particles is called __________ **soil**.

264. A layer of dead and decaying plant and animal remains is called __________ **Humus**.

265. __________ is called the uppermost layer of the soil and it is also known as 'A' horizon. **Topsoil**.

266. The removal of soil by running water and wind is known as soil __________ **Erosion**.

267. __________ soil is used to make pots, toys and statue - **Clayey**.

268. Nitrogen is absorbed by plant roots in the form of - **NO_3**.

269. Absorption of NH_4^+ by the root __________ the uptake of K^+ **reduces**.

270. The uptake of NH_4^+ by the plant roots result in the release of - **H^+**.

271. The conversion of organic form of an element into inorganic form is termed as **Mineralization**.

272. The conversion of amino acids into ammonium is called **Ammonification**.

273. C:N ratio which results in net mineralization is **20:1**.

274. The conversion of NH_4^+ into NO_3^- is called **Nitrification**.

275. About 80 % P is transported from soil to root surface through **Diffusion**.

276. In acidic soils, P availability decreases due to adsorption and precipitation with **Fe/Al oxides**.

277. If a bag of fertilizer is labeled "23-19-17", then it **contains 23% N, 19% K_2O, 17 % P_2O_5**.

278. An essential plant nutrient is defined as: **an element which a plant must have in order to complete its life cycle**.

279. Which three of the essential elements do plants get primarily from water and air? **O, H, C**.

280. Which three macronutrients are included in a so called "complete fertilizer? **N, K, P**.

281. Which of the following is not an essential plant nutrient? **Cadmium**.

282. DAP is a fertilizer source for **Nitrogen, Phosphorus**.

283. Phosphorus retention in soil is more if there is a dominance of **1:1 type minerals**.

284. There will be more fixations of P in soils having larger amounts of **clay**.

285. Phosphorus availability in most of the soils is in the pH range **5.5-6.5**.
286. The most commonly used salt for potassium determination in soil is **CH_3COONH_4**.
287. It is also termed as "muriate of potash" **KCl**.
288. Conversion of inorganic element into its organic form is termed as **Immobilization**.
289. Pakistani soils have average o.m content **< 1%**.
290. Nutrients which are required by plants in larger amounts are termed as **Macronutrients**.
291. A straight fertilizer is one having **only one nutrient**.
292. Nutrient ions present in soil reach to the root surface by **Diffusion**.
293. Delayed crop maturity is usually caused by too much **Nitrogen**.
294. The process whereby nitrate-N is reduced to nitrous oxide and elemental N is **Denitrification**.
295. Phosphorus fertility buildup factor values are usually low in **Wheat-Fallow**.
296. The loss of nitrogen as NH_3 gas from soil surface is termed **Volatilization**.
297. The high analysis N fertilizer is **Urea**.
298. The primary source of soil P is **Weathering of apatite minerals**.
299. To convert P to P_2O_5 multiply by **2.29**.
300. It is not part of any structural organic compound within plant body **K**.
301. Sulfate of potash has K_2O **50%**.
302. It is the central atom in the chlorophyll molecule **Mg**.
303. In flooded rice soils, best source of N is **Ammonium Sulfate, Urea, DAP**.
304. Volatilization losses of NH_3 are more when it is applied as **Broadcast Application**.
305. The most common method used in soil testing laboratories for P determination is **Olsen method**.
306. The most easily weatherable mineral which release K in soil is **Biotite**.
307. It can be replaced by Na in its osmoregulatory functions **K^+**.
308. They are structural component of chlorophyll molecule **N & Mg**.
309. Adequate supply of ___________ hastens maturity **Phosphorus**.

310. Its deficiency in plants causes chlorosis first in older leaves **Nitrogen**.

311. According to USDA classification, what is the diameter of medium sand particle - **0.25-0.50.**

312. The bulk density of sandy soil is about 1.6 g/cc. What is the approximate value of bulk density for organic matter - **0.4-0.6.**

313. The bulk density of a soil with particle density of 2.65gm/cc and pore space 40% would be: **1.59 gm/cc.**

314. The water which lies between wilting coefficient and field capacity is called available water. What is range of tension in hygroscopic coefficient? **31**.

315. What is the Clay percentage in sandy textural soil - **0-10.**

316. Pedology considers **soil** as a - **Natural body**.

317. A branch of science that deals with the study of origin of the soil, its **classification** and description is called as - **Pedology**.

318. Phosphorus uptake in alkali soil takes place in the form of ________ **PO_4^{3-}**.

319. Which **fertilizer** produce acidity in soil? Ammonium sulfate.

320. Maximum N content found in which of the "N fertilizer"? **Anhydrous ammonia.**

321. Urea **hydrolysis** in the soil produces: **Ammonium Carbonate**.

322. Rock phosphate has how much percentage of **P_2O_5 20-40%**.

323. Which of the following group of microbes increase phosphate solubility **Pseudomonas**.

324. The conversion factor for calculating P from **P_2O_5 x 0.44.**

325. The phenomenon slickenside found in which of the following soil? **Spodosol**.

326. Who is considered as father of soil testing - **Troug**.

327. Law of restitution was proposed by - **Liebig**.

328. Which horizon zone of Washing out - **A (Eluvial)**.

329. Which horizon zone of Washing in - **B (Illuvial)**.

330. Red soil mainly formed by which soil order - **Alfisols Soils**.

331. Black soil mainly formed by which soil order - **Vertisol soil**.

332. Largest Soil group is - **Alluvial soil group (143 mha)**.

333. C:N ratio of micro organism - **4:1 to 9:1**.

334. Equivalent acidity of Anhydrous ammonia - **148 meq/100 gm**.

335. Equivalent acidity of Urea - **80-85**.

336. Equivalent acidity of DAP - **77**.

337. "International potash of Institute" situated in - **Switzerland (1952)**.

338. Red colour of soil due to - **Hematite**.

339. Red yellow color of soil is associated with - **Ferric oxide**.

340. Which soil warmed up quickly - **Sandy soil**.

341. The B horizon is generally absent in which soil - **Black soil**.

342. Soil fertility and fertilizers was written by - **Samuet L. Tisdale**.

343. Which micro-nutrient are non-metal element - **Boron**.

344. Intracellular deposits of which element is known as plant opals - **Silicon**.

345. "Indian Society of soil science" (ISSS) established in which year - **Dec. 1934 (Kolkata)**.

346. "International Soil Reference and Information Center" situated in - **Netherland (1966)**.

347. "International Union of soil science" founded in - **1924 (Australia)**.

348. For nitrification, the optimum pH is - **8.5**.

349. White alkali soils are - **Saline soil**.

350. Humus fraction lightest in colour is - **Fulvic acid**.

351. Pedoturbation is the process of - **Mixing of the soil**.

352. In India, which soil order have negligible area - **Histosols**.

353. Particle density of clay soil is - **2.8**.

354. "Indian Institute of Soil Science" is situated in - **Bhopal (M.P.)**.

355. "Indian Institute of Soil and Water Conservation" is situated in - **Dehradun**.

356. First production of fertilizer in India - **Single Super phosphate (1906)**.

357. Calcareous soil is formed in arid as well as - **Humid region**.

358. Generally plant loses water through - **Transpiration**.

359. Active soil forming factor are - **Organism & climate**.

360. Which group of microorganism have add highest Biomass in soil - **Fungi**.

361. Soil color determination by - **Munsell's chart**.
362. Cation exchange is an important component of - **Soil fertility**.
363. Positive Bacteria color is - **Red**.
364. "Sulfur Institute" situated in - **Washington DC (1960)**.
365. First product of Urea if apply in soil - **Ammonium carbonate**.
366. Which is best method for application Fe and Zn - **Foliar application**.
367. Which is best suitable method for application Ca, B, S, and Mo - **Soil application**.
368. Kandic and Arfullic origin found in which order - **Alfisol**.
369. Surface color of Alkali soil is - **Black**.
370. Lowest category of soil classification includes - **Series**.
371. Sugarbeet is an indicator plant for - **Sodium**.
372. Wheat is an indicator plant for - **Copper**.
373. Chemical formula of siderite is **FeCo3**.
374. The origin of soil, its classification and its description are examined in **Pedology**.
375. ________ is study the various property of soil in relation to crop production. **Edaphology**.
376. The outer most 10 mile strata of the lithosphere are called the "________" **Earth crust**.
377. The solid zone of the earth sphere is the - **lithosphere**.
378. The fourth most abundant element in the earth crust is - **Iron**.
379. Most of the hard, naturally formed substance of the earth is referred to as **Rock**.
380. The solid zone of the earth's sphere is known as - **Lithosphere**.
381. The water forming zone of the earth's sphere is known as - **Hydrosphere**.
382. The author of "Introductory Soil Science" is - **D.K. Das**.
383. Science which deals with the study of the origin of rocks - **Petrology**.
384. Sedimentary rocks are also known as ________ rocks **Stratified**
385. Marble is metamorphic rock converted from limestone due to the action of ________ **Heat**.
386. Water plays an important role in the formation of ________ **Sedimentary**.

387. The individual grains sizes in rudaceous sedimentary rock are of - **boulders and pebbles**.

388. Rocks have individual grain size of clay **Argillaceous**.

389. Mineral is the source of P in soils **Apatite**.

390. The clay fraction of the soil particles has a diameter less than mm. **0.002**.

391. The minerals having sp.gravity < 2.85 are called minerals. **Light**.

392. __________ is a heavy mineral. **Pyrite**.

393. _________ is a primary mineral. **Feldspar**.

394. The chemical formula of hematite is __________ **Fe_2O_3**.

395. _______ when magma solidifiers at moderate depth is called **Intrusive**.

396. The native element of graphite is _______ **C**.

397. Rock is mixture is **Minerals**.

398. The native Element of diamond is ________ **Carbon**.

399. When the magma solidifies at great depth (2 to 3 miles) is called _____ **Plutonic**.

400. Physical weathering is also designated as _______ **Disintegration**.

401. Means chemical combination of water molecules with a particular mineral. **Hydration**.

402. May be considered as chemical principal agent of clay formation. **Hydrolysis**.

403. The material transported and deposited by water is ______ **Alluvium**.

404. The windblown materials are termed _______ **Loess**.

405. Is active soil forming factor. **Climate**.

406. The refers to the differences in elevation of the land surface. **Relief**.

407. The time that nature devotes to the formation of soil is termed as time. **Pedologic**.

408. Is the process of decomposition of organic matter and synthesis of new organic substances. **Humification**.

409. A soil profile contains _______ main horizons. **Four**.

410. The parent material from which the soil is formed is known as horizon. **C**.

411. Horizon ________ is called store house of plant nutrients. **B**.

412. Generally, particle density of normal soils is : **2.65 g/cc**.

413. The size of soil clay particle is ______ mm **<0.002**.

414. The attraction between clay and water particles is known as **adhesion**.

415. The bulk density of soil is always than particle density in cultivable soil. **less**.

416. Total porosity will be more in _______ **Clayey soil.** The attraction between similar particles is known as **Cohesion**.

417. The attraction between clay and water molecule is known as **Adhesion**.

418. Relative proportion of sand, silt and clay is referred as __________. **Soil texture**.

419. A simple method for the determination of moisture is **Neutron scattering**.

420. The zeta potential is the magnitude of ________ charge on colloidal particles. **Negative**.

421. The adsorption of phosphate ions is the ________ when the medium is neutral. **Lowest**.

422. Clay has a greater anion adsorbing capacity. **Kaolinitic**.

423. The minute colloidal clay particle is technically called **micelle**.

424. Ions behaves as polyvalent ions. **Hydrogen**.

425. The adsorption of phosphate ions is the lowest when the medium is **Neutral**.

426. Is example of the two type (1:1) of clay minerals. **Kaolinitic**.

427. First recognized cation exchange phenomenon. **Thompson**.

428. Millequivalent is defined as 1 milligram of or the amount of any other ions that will combine with or displace it. **Hydrogen**.

429. Water remains around the cation as a very thin film, which is called. **Water hull**.

430. Clay mineral has expanding type structure. **Montmorillonite**.

431. Particles less then mm size possess colloidal properties are known as soil colloids. **0.002**.

432. Is considered to the most important of the soils organisms bringing about the conversion of NO^2 - N to NO_3 - N. **Nitrobactor**.

433. Is considered to the most important of the soils organisms bringing about the conversion of $NH^4 \rightarrow NO^2$. **Nitrobactor**.

434. At point C:N ratio becomes more or less constant, stabilizes at **10:1**.

435. The conversion of organic phosphorus into available phosphorus. **Mineralisation**.

436. Is the scientific study of life in and on the soil. **Soil biology**.

437. The soil environment under the influence of plant roots is called the . **Rhizosphere**.

438. The chemical substances such as streptomycin a widly used antibiotic are obtained from. **soil fungus**.

439. Are transitional between bacteria and fungi. **Actinomycete**.

440. Contain chlorophyll and therefore, they manufacture their own food. **Algae**.

441. Regulating the number of the bacteria in the soil. **Actinomycetes**.

442. The process of conversion of proteins to amino acids is known as **aminization**.

443. Is non-symbiotic N-fixer in soil. **Rhizobium**.

444. Has directly functioned in nutrient uptake. **Mycorrhiza**.

445. Ion synthesis chlorophyll in plant. **Mg**.

446. Is a component of cell wall. **Calcium**.

447. % of the mineral nutrient supply is dissolved in soil water.

a) **< 0.20**

448. deficiency occurs as yellowing on older leaves. **N**.

449. per cent of mineral nutrients are bound in soil. **98**.

450. component remain as ash after combustion of dry plant material. **Mineral**.

451. *o*ccurs when there is concentration gradient of nutrients between root surface and surrounding soil solution **Diffusion**.

452. Is beneficial element. **Silicon**.

453. Zinc deficiency occurs on **Middle leaves**.

454. Among the following which is a micronutrient element for higher plant. **Zinc**.

455. Both clay minerals and humic colloids have a ______ net charge so that they attract and adsorb primarily cations in soils. **Negative**.

456. Cauliflower is an indicator plants for deficiencies. **Mo**.

457. Clay having ______ charge so that they attract and adsorb primary cations. **Negative**.

458. Contact exchange theory is highly favourable for the uptake **Phosphorus**.

459. Deficiency of _________ appears as brown spot on paddy. **Zn**.

460. Deficiency of ________ shows marginal burning of leaves, irregular fruit development of plant - **K**.

461. Deficiency of ______ appears shortening of internodes in plants. **Zinc**.

462. Essentiality criteria of nutrients are given by **Arnon**.

463. Highly mobile nutrient in plant is ________ **Nitrogen**.

464. Immobile nutrient in plant is __________ **Calcium**.

465. Inter veinal yellowing or motling of young leaves due to deficiency in plant. **Mn**.

466. Khaira disease of rice is caused by ________ deficiency. **Zinc**.

467. Movement of nutrient ions and salts along with moving water is terms as **Mass flow**.

468. Yellowing of new leaves occur due to deficiency of **sulphur**.

469. The ______ is an example of non – metal nutrient **Mo**.

470. Yellowing of new leaves occur due to deficiency of **sulphur**.

471. The author of "Soil Fertility Theory & Practice" is ______ **Kanwar**.

472. The average concentration of nitrogen (%) in plant tissue is. **1.5**.

473. Whiptail in cauliflower is due to the deficiency of **Mo**.

474. The element which is immobile in the plant system is **Boron**.

475. Total soluble salts of sodic soil is **<0.1%**.

476. Which is not a beneficial element for the plant is **Iron**.

477. Can grow well under very acid soil and caused diseases like root rot of Tobacco. **Fungi**.

478. Soil are formed in regions of intensive rainfall and leaching. **Acid**.

479. Is a high acid tolerant crop. **Rice**.

480. Is used for reclamation of acidic soil. **Lime**.

481. Weed has been found very suitable for reclamation of alkali soil. ***Argemonemexican***.

482. Is sensitive to 10-20 ESP levels in soil **beans**.

483. Is also known as alkali soil. **Black alkali**.

484. Is the most suitable green manuring crop for the reclamation of alkali soil. **Dhaincha**.

485. is also known as saline soil. **White alkali**.

486. 1 me gypsum ($CaSO_4 \cdot 2H_2O$)/100 g = _______ mg/100 g gypsum. **86.**

487. 1me Ca =mg Ca **20**.

488. 1me Na =mg Na **23**.

489. Acidic soils develop under areas **High rainfall**.

490. Among the following cations which have highest tendency to adsorption with soil colloids. **Al**+3.

491. Availability of ______ increases in acidic soils. **Manganese**.

492. is example of the mostpersistent herbicides **Atrazine**.

493. Calcareous soils are mostly developed under ______ climate. **Arid & semi-arid**.

494. Dominant cation in alkali soil is. **Sodium Chloride**.

495. Exchangeable sodium percent (ESP) less than ______ is safe for tolerable soil physical condition. **15**.

496. For the replacement of 1 me of exch. Na, it requires _______ kg gypsum/ ha. **1926.4**.

497. Generally the osmotic pressure of saline soil is. **High**.

498. Generally the physical condition of sodic soil is **deflocculated**.

499. Generally % of GR (Schoonover's method) has been found most satisfactory in many type of soils **50-75**.

500. Gypsum equivalent of sulphur is. **0.186**.

501. Glacial soils are found in-Utrakhand.

502. According to land capability claassification class-I is very good land with-no limitations.

503. Dolomite is rich in-Mg.

4

Entomology

1. Who was the first Entomologist to the Government of India? **Lionel de Niceville**.
2. Who succeeded Lefroy and devoted his attention mainly to insect taxonomy and built up an excellent library and a valuable insect collection? **T.B. Fletcher**.
3. When was the Zoological Survey of India started in the Indian Museum at Calcutta? **1916**.
4. Who was the first Plant Protection Adviser to Government of India? **Dr. Hem Singh Pruthi**.
5. When the Locust Warning Organization was established in India? **1939**.
6. Who was first Imperial Entomologist? **Maxwell Lefroy**.
7. In which insect cerci is modified as forceps? **Earwig**.
8. Name insect orders having abdomen with filamentous cerci. **Plecoptera**.
9. Which insect is biggest in insect world in term of total wing surface area? **Atlas moth (*Attacusatlas*)**.
10. In which insect right mandible is reduced in size? **Thrips**.
11. Which is fastest flying insect? **Dragonfly**.
12. It is the largest order in the animal kingdom. **Coleopteran**.
13. Human disease pathogens are transmitted by insects of which order? **Diptera**.
14. A fatal disease bubonic plague is transmitted by __________ **Fleas**.
15. Which insect order(s) has/have both harmful and beneficial insects? **Colleoptera, Diptera & Hymenoptera**.
16. All arthropods have **Segmented bodies, Jointed appendages and Bilateral symmetry**.

17. Exoskeleton provides: **Muscle attachement, Good protection & Water conservation**.
18. Entomological society of India was established in the year **1937**.
19. Honey bees are more closely related to **Ants**.
20. House flies are more closely related to **Mosquitoes**.
21. Sawflies are more closely related to **Wasp**.
22. ________ is/are common in orders Protura, Diplura and Thysanura? **Winglessness & Two pair of wings**.
23. In which class the body is divided into cephalothorax and abdomen? **Arachnida**.
24. Sugarcane pyrilla belong to family **Fulgoridae**.
25. According to A. D. Imms insect is divided into how many order? **29**.
26. Insects of which order possess scale wings? **Lepidoptera**.
27. Silver fish belonging to the order **Protura**.
28. Silphidae and Scarabaeidae are types of **Beetles**.
29. To which class of arthropods do lobsters and shrimp belong? **Crustacea**.
30. Which arthropods have chelicerae? **Spiders**.
31. Orthopteran insects should be pin through the **Right posterior quarter of Prothorax**.
32. For dry storage of coleopteran insects, it should be pin through the **Right elytra near its front**.
33. Shield bug should be pin through the **Scutellum**.
34. Completely parasitic order is **Strepsiptera**.
35. From the following which insect's name is wrongly written? **Mango fruitfly**.
36. From the following which is not dipteran fly? **Mustard Sawfly**.
37. From the following which is true fly? **Tachinid fly**.
38. The theory and practice of identifying, describing naming and classifying organisms is called **Taxonomy**.
39. The study of the kinds and diversity of organisms and the relationships among them is called **Systematics**.
40. Insects of which order is only phytophagous? **Phasmida**.
41. The level of taxonomy concerned with the arrangement of species in the biological hierarchy is **Beta-taxonomy**.

42. Different names given to the same taxon are called **Synonym**.
43. Gelechidae is the family of which order? **Lepidoptera**.
44. It is not a family of order hymenoptera. **Aphididae**.
45. Insect order has the great influence in biological control component **Hymenoptera**.
46. Which insects should be submitted in a vial of alcohol? **All soft-bodied insects**.
47. Which order is not aquatic order? **Lepidoptera**.
48. All ametabolous insects are: **Wingless**.
49. Which insect order is most commonly found in soil litter? **Collembola**.
50. Hermaproditism is very common in ***Icerya purchase***.
51. Indian lac Research Institute is situated at - **Ranchi**..
52. The part of insect body is **Head, thorax and abdomen**.
53. Which of the following insect have prognathous type of head? **Termites**.
54. In a mandible, the distal biting surface having pointed projections is called as **Incisor cusps**.
55. The legs of insects arises from **Thorax**.
56. The mouth parts situated on the **Head region**.
57. The antennae is situated at **Head**.
58. How many parts are there in the insect's antennae? **3 parts**.
59. How many segment insect legs have? **5 segments**.
60. The anal cerci is situated on the **Abdomen**.
61. The thorax of insect is **3 segmented**.
62. The insects head is composed of embryonic segments **6**.
63. How many pairs of antennae insects have? **1 pairs**.
64. The body wall of insect is composed of **3 layers**.
65. The outer most layer of insect body is called **Epidermis or Hypodermis**.
66. The epicuticle of insect is composed of **4 layers**.
67. The cuticular layer is the part of **Epicuticle**.
68. Polyphenol layer is the part of **Epicuticle**.
69. Cement layer is the part of **Epicuticle**.
70. The cuticle of insect is **Cellular**.

71. Monocondylic single mandible is found in **Thysanoptera**.
72. The male cockroach can be distinguished from the female by having **Anal styles**.
73. How many axillary sclerites are present in insects? **3**.
74. How many axillary sclerites are found in the order hymenoptera and orthoptera? **Four**.
75. In which insect order, axillary sclerites are absent **Odonata & Ephemeroptera**.
76. The body of insect is covered with hard plates which are known as **Sclerites**.
77. The sclerites are attached to each other with a thin membrane is called **Suture**.
78. The joints of grooves are called **Suture**.
79. The basement membrane is the **Inner middle**.
80. The ocellary segment related to the **Head**.
81. Intercalary segment is related to the **Head**.
82. How many types of head found in insects **Pro, Hypo and Opisthognathous**.
83. In grasshopper which type of head is found? **Hypognathous**.
84. In Pyrilla which type of head is found? **Prognathous**.
85. Which type of head found in weevils? **Prognathous**.
86. In Hemipterous (Bugs) order which type of head is found? **Opisthognathous**.
87. Frons related to the **Thorax**.
88. Gena related to the **Thorax**.
89. Clypeus is situated at **Thorax**.
90. What is the shape of the epicranial suture? **Inverted Y shaped**.
91. The Johnston's organ works as **Auditory organ**.
92. Setacious type antennae found in **Cockroach**.
93. Moniliform type antennae found in **Termite**.
94. Bipectinate type antennae found in **Silk moth**.
95. Serrate type antennae found in **Pulse beetle**.
96. Flabellate type antennae found in **Cedar beetle**.

97. Clavate type antennae found in **Butter fly**.
98. The lower lip of insect is known as **Labium**.
99. Insect tongue is **Hypopharyn**.
100. The labrum attached to the **Clypeus**.
101. How many pairs of mandible and maxilla are found in insects? **One pairs**.
102. Cardo is the part of **Maxilla**.
103. The maxillary palps are arises from **Stypes**.
104. The basal segment of maxillary palps is known as **Palpifer**.
105. How many segment are usually found in maxillary palp? **5**.
106. Galea is the part of **Maxilla**.
107. Sub mentum and Mentum is the part of **Labium**.
108. How many pairs of labial palps are found in mentum? **One pair**.
109. The labial palpi is attached to lateral sclerite known as **Palpiger**.
110. Median pair of lobes of pre mentumis known as **Glossa**.
111. Lateral pair of lobes of pre mentum is known as **Para glossa**.
112. Vestigial wings are found in ***Blatta sp***.
113. In piercing and sucking type of mouth parts labium modified into **Rostrum**.
114. Which type of head is found in the Sucking type insects? **Opisthognathous**.
115. The epistomal suture or fronto clypeal suture found between **Frons and clypeus**.
116. The clypeo-labral suture is found between **Clypeus-labrum**.
117. What is the position of labial palp in piercing and sucking type of mouth part? **Absent**.
118. The mandible and maxilla of bugs modified into long slender sclerotized hair like structure is called? **Stylet**.
119. Palpiger is the sensory part of: **Labium**.
120. Which part of the maxillae is modified from a coiled spring in siphoning type of mouth parts? **Galea**.
121. The wings of insects arises from **Abdomen**.
122. Wings arises jointly on **Pterothorax**.
123. The insect thorax is _________ segmented **Three**.
124. Which type of leg is primarily designed to capture and hold prey? **Raptorial**.

125. How many legs Insects have? **3 pairs**.
126. Which structure is helpful to housefly for walking on smooth surface? **Pulvilli**.
127. Which type of legs are found in mole cricket **Fossorial**.
128. Which type of forelegs is found in preying mantid? **Raptorial**.
129. The function of middle legs of honey bees is **Pollen brushing**.
130. In which insect prothorax and metathorax are reduced and fused with well developed mesothorax? **House fly**.
131. In which insect only thorax and abdomen visible dorsally? **Lesser grain borer**.
132. Firefly emits bright light from **Either side of thorax**.
133. In which insect, Scantorial type of leg is found? **Louse**.
134. Meso and metathorax is collectively known as _______ **Pterothorax**.
135. The forewing of coleopteran insects modified in to for protection of membranous hind wings. **Elytra**.
136. In house fly, hind wings are reduced to slender knob and a stalk structure known as **Halters**.
137. In honey bee workers, the pollen basket is found in **Tibia**.
138. Adult mite have ________ **Four pair of legs**.
139. Which is primitive wingless insect order? **Thysanura**.
140. Tympanum organ is related to the **Sound producing**.
141. Which structure is always associated with the Hymenoptera? **Hamuli**.
142. Minute spines which form a pattern on the extremity of false leg of lepidopterous larvae is called **Chrochets**.
143. Hemiptera and Hymenoptera are similar because both have: **Neopterous wings**.
144. Which caste of honey bee has foragial type of leg? **Worker**.
145. In case of loopers how many pairs of prolegs are present? **2**.
146. Structure that help the cockroach to walk on smooth surface is **Plantulae**.
147. In which insect order there is no functional leg? **Strepsiptera**.
148. In dragonflies and damselfly, there is an opaque-spot near the costal margin of wing called **Pterostigma**.

149. Wings are leathery or parchment like are called **Tegmina**.
150. The basal half of the wing is thick and leathery with distal half is membranous is known as **Hemielytra**.
151. Which type of wings are found in beetles and weevils? **Elytra**.
152. The legs of insects arises from **Thorax**.
153. The wings of insects arises from **Abdomen**.
154. The male genital appendages are situated at abdominal segment **9th**.
155. In the female genital appendages are situated at abdominal segment **8th and 9th**.
156. The main role of Ovipositor is **To help in egg laying**.
157. Aedeagus is also know as **Penis**.
158. The female genital chamber is often called **Vagina**.
159. What is the main function of insect abdomen? **Storage and reproduction**.
160. Cornicles are found in **Aphids**.
161. The last segment in the abdomen of cockroach in addition of cerci will have a pair of **Anal styli**.
162. The meaning of term "Propodium" is **Fused segments of the abdomen**.
163. Glow worms (*Lamprophoroustradus*) emits bright light from the luminous organs present on **Hind segments of abdomen**.
164. Prolegs are the appendages of **Abdomen**.
165. The alimentary canal of insect is divided into parts. **3**.
166. Fore gut is in origin. **Ectoderm**.
167. Crop and gizzard is a part of **Foregut**.
168. Absorption of digested food takes place in **Midgut**.
169. In insect excretion is take place by **Malpighian tubules**.
170. Main excretory material in insect is **Uric acid**.
171. The body cavity through which blood flow is called **Haemocoel**.
172. Major blood sugar is **Trehalose**.
173. The insect circulatory system is which type **Open**.
174. Nerve cord in insect is located **Ventrally**.
175. Supra oesophagal ganglion is also called as **Brain**.
176. Spiracles are also know as **Stigmata**.

177. Holopneustic respiration refers to **All spiracles open**.
178. In Arachnida respiration takes place by **Trachea**.
179. The digestion of cellulose in termite is done by **Protozoan flora**.
180. The inorganic iron are absorbed in to **Rectum**.
181. Afferent neuron is also knows as **Sensory neuron**.
182. Efferent neuron is also knows as **Motor neuron**.
183. Ventral nerve cord is the part of **Nervous system**.
184. In cockroach ootheca is secreted by **Colleteral gland**.
185. Reproduction in larval or other pre adult form is known as **Paedogenesis**.
186. The light is produced by the oxidation of **Leuciiferin**.
187. Cantharidin is obtained from **Blister beetle**.
188. Insect nervous system is **Chlorigenic**.
189. Neuropile is **Mass of axon**.
190. The type of parthenogenesis in which only female are produce **Thelytoky**.
191. The function of spermatheca is **Seminal vesicle**.
192. Larvae of mosquito is called **Wriggler**.
193. The interval between the moulting is knows as **Stadium**.
194. The separation of old cuticle and epidermis is called **Apolysis**.
195. The Moulting fluid contains **Proteinase and Chitinase**.
196. The larvae of white grub are **Scarabeiform**.
197. Caterpillar are _______ **Polypod**.
198. The pupa of butterfly is called **Chrysalis**.
199. The pupa of mosquito is called **Tumbler**.
200. Which type of pupa found in moths? **Obtect**.
201. Aquatic nymph is called **Naiad**.
202. The cast cuticle or skin during mouting is called **Exuviae**.
203. Insect with complete metamorphosis the immature are called **Larva**.
204. In exopterygota insect order which type life cylcle is found **Incomplete**.
205. Complete life cycle how many stage are found **4**.
206. Incomplete life cycle how many stage are found **3**.
207. Book "Elements of Economic Entomology" written by **David and Kumarswami**.

208. Book "Integrated Pest Management, Concepts and Approaches" is written by **G.S. Dhaliwal & Arora**.

209. Book "Pesticides Application Equipment" is written by **O.S. Bindra and H. Singh**.

210. Book "Introduction of Insect Pest Management" is written by **Metcalf R.L. & Luckman W.H.**

211. Book "Biological Control of Insect Pests" is written by **Saxena A.B.**

212. Book "Biological Control of Insect Pests and Weeds" is written by **De Bach P.**

213. Book "Principles and Procedures of Plant Protection" is written by **S.B. Chattopadhyay**.

214. Which one is monophagous pest **Tur pod fly**.

215. Which one is International pest ? **Locust**.

216. Which pest is polyphagous in nature **Helicoverpa**.

217. Which one is national pest ? **Helicoverpa**.

218. Which one is oligophagous pest ? ***E. vittella***.

219. The control measures adopted by man are called as **Applied control**.

220. The control of insects by topographic factors is called as **Cultural control**.

221. The control measures adopted adopted after appearance of pest are called as **Curative control**.

222. The control measures adopted adopted before appearance of pest are called as **Prophylactic control**.

223. Earthing up one or two months after planting minimize sugarcane **Early shoot borer**.

224. High seed rate of sorghum is recommended for the control of **Shootfly**.

225. Growing of tomato as trap crop in citrus for the management of **Fruit sucking moth**.

226. Adoption of appropriate farm practices to eliminate or reduce the pest population is called **Cultural**.

227. Clipping-off tips of rice seedlings before planting eliminate egg masses of **Rice Stem borer**.

228. Crop rotation of cotton after lady's finger increased population of **Earias vittella**.

229. Castor should be sown around 15th-30th August to manage the pest **Semi looper**.
230. Mustard crop should be sown between 15 to 25 October to protect the crop against **Aphids**.
231. Host evasion is also known as **Pseudo resistance**.
232. Early sowing of sorghum immediately after onset of monsoon is useful to escape attack of **Shoot fly**.
233. Castor should be sown between 15 to 30 August for management of **semi-looper**.
234. Small planting of a susceptible or preferred crop around/in- between major crop is called as **Trap crop**.
235. Castor grown around cotton is trap crop for ***Spodoptera***.
236. Castor grown around chilli is trap crop for **fruit borer**.
237. Tomato grown between citrus is trap crop for **fruit sucking moth**.
238. Prunning of dried twigs to eliminate pest like **scale insects**.
239. Fertilizer application with Super phosphate will reduce attack of **rice weevil**.
240. Flooding rice nurseries to eliminate the attack of **Armyworm**.
241. Tar application on stem of trees is to prevent attack of **Rat**.
242. Light trap is used to attract the pest **Nocturnal pest**.
243. Brushing of Woolen fabrics is useful for control **carpet beetle**.
244. Hopper dozer is used to Kill nymphs of **locusts**.
245. Alkathene bands with double girdle is useful to prevent **Mango mealy bug**.
246. Bagging fruits of pomegranate is useful to protect fruits from **Anar butterfly**.
247. Barriers are used against **Nonflying insects**.
248. Sand barriers are used for protecting **Stored grains**.
249. Water barrier is used against **Ant**.
250. Cotton seed exposed to 56 °C for 4 hrs. And 15 minutes completely killed hidden infestation of pink boll worm **Pink boll worm**.
251. Alternate drying and wetting rice fields is useful against **BPH**.
252. Providing light in storage godowns as the lighting reduces the fertility of **Indian meal moth**.

253. Pest that introduced accidentally into India is **Pink bollworm**.
254. Pest that introduced accidentally into India is **Cottony cushion scale**.
255. Pest that introduced accidentally into India is **Apple Wooly aphid**.
256. Pest that introduced accidentally into India is **San Jose scale**.
257. Pest that introduced accidentally into India is **Potato tuber moth**.
258. Pest that introduced accidentally into India is **Golden nematode**.
259. Pest that introduced accidentally into India is **Subabul psyllid**.
260. Management of insects by predator and parasite is known as **Biological control**.
261. *Trichogramma* is parasitize on **Egg**.
262. Those parasite which parasitise the Larva are known as **Larval Parasite**.
263. Which one is a beneficial insect **Honeybee**.
264. *Trichogramma minutes* is **Egg parasite**.
265. *Epiricania melanoluca* is **Adult parasite**.
266. *Beauveria bassiana* is **Entomopathogen ic fungi**.
267. *Steinernema carpocapsae* is **Entomopathoge nic nematode**.
268. A chemical that prevents feeding of insect on a treated material without necessarily killing it is known as **Antifeedant**.
269. Technical name of Marshal is **Carbosulfan**.
270. Trade name of Chlorpyriphos is **Dursban**.
271. Technical name of Regent is **Fipronil**.
272. Nicotine derives from leaves of ***Nicotiana tabacum***.
273. Trade name of Dimethoate is **Rogor**.
274. Technical name of Curacron is **Profenofos**.
275. Trade name of monocrotophos is **Nuvacron**.
276. Trade name of triazophos is **Hostathion**.
277. Which one is systemic insecticide ? **Dimethoate**.
278. Spinosad prepared from ***Saccharopolyspora spinosa***.
279. Nicotine is derived from ______ ***Nicotiana tabacum***.
280. Which of the following compound is a plant origin? **Nicotine**.
281. ________ is belongs to organochlorine group? **BHC**.

282. ________ is trade name of endosulfan? **Thiodan 35% EC**.
283. ________ is a technical name of lindane 0.65% dust & 10% G? **BHC**.
284. ________ of the following organochlorine compound is phytotoxic to cucurbit? **DDT**.
285. Mode of entry of organochlorine compound is **Stomach & Contact**.
286. Technical name of DDVP 76% EC is ______ **Dichlorovos**.
287. ________ is trade name of monocrotophos? **Nuvecron 36% EC**.
288. ________ is trade name of phosphamidon? **Dimecron 85% EC**.
289. ________ is trade name of methyl parathion? **Metacid 50% EC**.
290. ________ is trade name of fenitrothion? **Folithion 50% EC**.
291. ________ is trade name of fenthion? **Leybacid, 100% EC**.
292. ________ is trade name of chlorpyriphos? **Dursban 20% EC**.
293. ________ is trade name of diazinon? **Basudin 20% EC**.
294. ________ is belonging to organophosphorus group? **Triazophos**.
295. Mode of entry of monocrotophos is **Systemic**.
296. Mode of entry of methyl parathion is _____ **Contact & Stomach**.
297. Mode of entry of fenitrothion is**Contact & Stomach**.
298. Mode of entry of chlorpyriphos is ______ **Contact & Stomach**.
299. Mode of entry of diazinon is _______ **Contact & Stomach**.
300. Mode of entry of methyl-o- demeton is**Contact & Stomach**.
301. Mode of entry of quinalphos is _________ **Contact**.
302. Mode of entry of phorate is **Systemic**.
303. Mode of entry of thiometon is **Systemic**.
304. ________ is trade name of carbaryl **Sevin 50% WP**.
305. ________ is trade name of carbofuran? **Furadan 3% G**.
306. ________ is trade name of aldicarb? **Temic 10% G**.
307. ________ is trade name of indoxacarb? **Avaunt 14.5 SC**.
308. ________ is trade name of thiocarb? **Larvin 75 WP**.
309. ________ is trade name of carbosulfan? **Marshal 25 EC**.
310. Mode of entry of carbaryl is **Contact**.
311. Mode of entry of indoxacarb is **Contact & Stomach**.

312. ____________ compound is synthetic pyrethroid? **Fluvalinate**.
313. ________ is a trade name of permathrin? **Permasect 10 & 20% EC**.
314. __________ is a trade name of fenvalate? **Sumicidin 20%**.
315. ____________ is a trade name of decamethrin? **Decis2.8% EC**.
316. ________ is a trade name of beta cyfluthrin? **Bulldock 0.25 SL**.
317. ____________ is a trade name of fluvalinate? **Mavrik 20EC**.
318. Mode of entry of permathrin is _____ **Contact & stomach**.
319. Mode of entry of cypermathrin is _____ **Contact**.
320. Mode of entry of fenvalate is ______ **Contact**.
321. Mode of entry of decamethrin is ________ **Contact**.
322. Mode of entry of lamda cyhalothrin is ________ **Contact & stomach**.
323. Mode of entry of beta cyfluthrin is ______ **Contact & stomach**.
324. Mode of entry of fluvalinate is **Contact & stomach**.
325. ______ is a trade name of acetamiprid? **Pride®**.
326. Mode of entry of imidacloprid is ______ **Systemic**.
327. Mode of entry of acetamiprid is **Systemic**.
328. Avermectin is derived from ______ ***Streptomyces avermetilis***.
329. Spinosyns is derived from ______ ***Saccharopolys pora spinosa***.
330. Cartap hydrochloride is derived from _________ ***Lumbriconereis heteropoda***.
331. Avermectins is commercially available as a ______ **Abamectin**.
332. Spinosyns is commercially available as a ______ **Spinosad**.
333. Cartap hydrochloride is extracted from **Marine annelid**.
334. Cartap hydrochloride is commercially available as______ **Caldan 50SP**.
335. ______ is a trade name of fipronil? **Regent®**.
336. ______ is a trade name of indoxacarb? **Avaunt 14.5 SC**.
337. Mode of entry of avermectins is______ **Systemic**.
338. Mode of entry of spinosyns is ______ **Contact & stomach**.
339. Imidacloprid is effective against ______ **Sucking pest**.
340. The abamectin is commercially available in market under trade name **Emamectin benzoate**.

341. The acemethrin is commercially available in market under trade name **Chinmix**.

342. The alanicarb is commercially available in market under trade name **Onic**.

343. The allethrin is commercially available in market under trade name **Pynamin**.

344. The acetamiprid is commercially available in market under trade name **Mospilon**.

345. The *Beauveria bassiana* is commercially available in market under trade name **Biosoft**.

346. Which one is abiotic factor? **Temperature**.

347. Which one is biotic factor? **Parasite**.

348. The sphere of life is also described as **Biosphere**.

349. A relatively artificial ecosystem in an agricultural field, pasture, or orchard is known as ______ **Agro-ecosystem**.

350. The optimum temperature for most of the insect species range between **13 to 40°C**.

351. The body condition of all insects consists mostly of water & they contain water. **80-90%**.

352. The number of hours of light in a day length (24 hours) termed as ______ **Photoperiodism**.

353. Pest cause 5-10% damage is known as **Minor pest**.

354. Pest cause > 10% damage is known as **Major pest**.

355. Insect causes a loss of < 5% is considered as **Negligible pest**.

356. The period between female adult emergence and first egg laying is known as **Pre oviposition period**.

357. The main objective of pest management is ______ **Reduce pest status to ETL**.

358. Adoption of appropriate farm practices to eliminate or reduce the pest population is called **Cultural**.

359. Regular removal of weeds, grasses & pests affected parts of plant is called as **Cultural control**.

360. Use of pest free healthy seeds/planting material is called as **Cultural control**.

361. Treating the seeds before sowing is called as **Cultural control**.

362. Periodic racking up of manure pits prevent breeding of rhinoceros beetle is **Cultural control**.
363. Manipulation of cultural practices to the disadvantage of pests is called as **Cultural control**.
364. Coconut trees protected with tin band to prevent damage of **Rat**.
365. Dip the infested fruits in 50° C + 0.5 °C water for 2 hrs - kill all the stages. **Mango stone weevil**.
366. Tar application on stem of trees is to prevent attack of **Rat**.
367. Hooking iron hook in tree trunk is used against **Rhinoceros beetle**.
368. Crushing Bed bugs and lice is called as **Mechanical control**.
369. Delousing method for Head louse is **Mechanical control**.
370. Brushing of Woolen fabrics is useful for control **carpet beetle**.
371. Hopper dozer is used to Kill nymphs of **locusts**.
372. Hot water treatment (50-55°C for 15 min) is useful to manage **Rice white tip nematode**.
373. Dip the infested fruits of Mango in 50° C + 0.5 °C water for 2 hrs to kill all the stages of **Mango stone weevil**.
374. Flame throwers are used against **locusts**.
375. Cold storage of fruits and vegetables are useful to **kill fruit flies**.
376. Foreign pests from which India is free **Mediterranean fruitfly**.
377. Foreign pests from which India is free **Cotton boll weevil**.
378. Foreign pests from which India is free **Codling moth of apple**.
379. Mantid is a **Predator**.
380. Ladybird beetle is a **Predator**.
381. Chrysopa is a **Predator**.
382. Spider is a **Predator**.
383. Syrphid fly is a **Predator**.
384. Helilure attract the male of ***Helicoverpa***.
385. Spodolure attract the male of ***Spodoptera***.
386. Rotenon is derived from _________ ***Derris elliptica***.
387. Sabadilla is derived from _________ ***Schoenocaulon otticinale***.
388. Tephrosin is derived from _________ ***Tephrosia vogelli***.

389. Ryania is derived from _________ ***Ryania speciosa***.
390. Rotenon is act as a _________ **Fish poison**
391. Sabadilla act as a ______ **Contact poison**.
392. Relationships between organisms and their environment is called **Ecology**.
393. Dormancy in insect during winter is called as **Hibernation**.
394. Dormancy in insect during summer is called as **Aestivation**.
395. Study of interactions among species, ecological succession is known as _______ **Community Ecology**.
396. Study of population growth and factors that affect growth is known as _________ **Population Ecology**.
397. A relatively artificial ecosystem in an agricultural field, pasture, or orchard is known as _________ **Agro-ecosystem**.
398. The number of insects per unit (of measure) is known as _________ **Density**.
399. Physiological state of arrested metabolism, growth, and development is known as _________ **Diapause**.
400. A disease outbreak within an insect population is called _____ **Epizootic**.
401. A group of individuals similar in structure and capable of interbreeding is known as **Species**.
402. Insect occur in a few isolated localities is known as **Sporadic pest**.
403. The lowest number of insects that will cause economic damage is ______ **EIL**.
404. It is the maximum level of pest population that will not cause the economic damage to crop ________ **Economic Threshhold Level**.
405. Certain insect occur more frequently on a crop which have a close association with a particular crop **Regular pest**.
406. When the infestation occurs in a regular feature and confined mostly to a particular area or locality it is known as **Endemic pest**.
407. Mustard sown 15th to 25th October for the management of **Aphid**.
408. Trap crop of castor in or around cotton crop is beneficial to control ***Spodoptera***.
409. Growing of marigold as trap crop in chickpea for the management of ***Helicoverpa***.
410. Crop rotation of cotton after lady's finger increased population of ***Earias vittella.*** Castor should be sown around 15th -30th August to manage the pest **Semi looper**.

411. Mustard crop should be sown between 15 to 25 October to protect the crop against **Aphids**.

412. Host evasion is also known as **Pseudo resistance**.

413. Castor should be sown between 15 to 30 August for management of **semi-looper**.

414. Marigold grown around chickpea is trap crop for ***Helicoverpa***.

415. Yellow sticky trap is used to attract the pest **Aphids**.

416. Light trap is used to attract the pest **Nocturnal pest**.

417. Sun drying of storage grains is known as **Physical control**.

418. Hand picking of insect control is known as **Mechanical control**.

419. Operaton by machinary or manual force is known as **Mechanical control**.

420. Beating or Swatting housefly and mosquito is **Mechanical control**.

421. Shaking neem tree to dislodge white grub beetles is **Mechanical control**.

422. Crushing Bed bugs and lice is called as **Mechanical control**.

423. Delousing method for Head louse is **Mechanical control**.

424. Gamma irradiation from Co_{60} is used to **Sterilize the insects**.

425. Gamma irradiation from Co_{60} is used to sterilize the insect **cattle screw worm fly**.

426. Male sterile technique in cattle screw worm flywas discovered by. **E.F. Knipling**.

427. Treating the stored pulses with vegetable oils to prevent the oviposition of **Pulse beetle**.

428. Yellow colour is preferred by **aphids**.

429. Red earth treatment to red gram is applied to make Injury to **wax layer**.

430. Preventing the entry and establishment of foreign plant pest in a country through enactment is called **Legislative Control**.

431. Destructive Insects and Pests Act- of India (DIPA) enacted in **1914**.

432. A certificate issued by State Entomologist and Pathologists that the plant or seed material is free from any pest is known as **Phytosanitary**.

433. The First Plant Protection Quarantine Act was passed in the year **1905**.

434. US Plant Quarantine Act **1912**.

435. Madras Agricultural Pests and Diseases Act **1919**.

436. The First Plant Protection Quarantine Act was passed in the year **1912**.

437. The Insecticides Act **1968**.

438. Management of insects by predator and parasite is known as **Biological control**.

439. *Bacillus papillae* is effective against **White grub**.

440. The larva of blister beetle is feeds upon the eggs of **Grass hopper**.

441. If a parasites is contained to a single species of insect it is called **Monophagous**.

442. Parasite is capable to develop upon a few closely related host species is called **Oligophagous**.

443. Parasite can develop upon a number of widely different host species is called **Polyphagous**.

444. When parasites attack on the host directly is known as **Primary parasites**.

445. When a parasite attacks on the secondary parasites is called as **Tartiary parasites**.

446. Attack of a parasite on an insect which is already parasite of other insect is **Hyper paeasitism**.

447. When many individuals of the same species of the parasite attack a single host **Super parasitism**.

448. When more than one species of parasites present on the same host is **Multiparasitism**.

449. Those parasitise the host larvae and continue their development by remaining on external surface of host is called **Ecto parasite**.

450. Those parasitise the host larvae and continue their development by remaining in host body is called **Endo parasite**.

451. Those which parasitise the host pupae and continue their development within the host pupae is **Pupal parasite**.

452. Those prasitise the host adult stage is known as **Adult parasite**.

453 _________ is parasite of black coconut headed caterpillar, *Nephantis serinopa*. ***Apanteles flavipes***.

454 _________ is parasite of black coconut headed caterpillar, *Nephantis serinopa*. ***Bracon brevicornis***.

455. ____ is parasite of rice stem borer, *Scirtophaga nivella* ***Isotima javensis***.

456. _______ is natural enemy of cottony cushion scale ***Rodolia cardinalis***

457. ________ is natural enemy of *Eriosoma lanigera* ***Aphelinus mali***.

458. ______ is natural enemy of Sugarcane shoot borer ***Trichogramma australicum***.

459. Nematodes specially rhabditids are found to have symbiotic relationship with the **Bacteria**.

460. DD-136 is a product made of **Nematode**.

461. *Bacillus thuringiensis* is type bacteria **Spore forming**.

462. NPV is considered as **Stomach poison**.

463. *Beauveria bassiana* is considered as **Contact poison**.

464. _________ NPV causes adverse effect on **Cell nuclei**.

465. CPV causes adverse effect on **Cell cytoplasm**.

466. *Nosema bombycis* is a, which cause adverse effect on silk worms **Protozoan**.

467. A chemical that attract the insect on a treated material without necessarily killing it is known as **Attractant**.

468. Oil of Citronella act as repellent for **Mosquito**.

469. Methyl eugenol act as attractant for **Fruitfly**.

470. Lethal concentration required for killing 50% of test insect known as $\mathbf{LC_{50}}$.

471. Lethal dose required for killing 50% of test insect known as $\mathbf{LD_{50}}$.

472. Lethal dose required for killing 50% of insect and it is expressed as ________ $\mathbf{LD_{50}}$

473. Lethal concentration required for killing 90% of insect and it is expressed as _____ $\mathbf{LC_{90}}$.

474. The median knockdown time required to kill 50 per cent of test animals is expressed as _____ $\mathbf{KT_{50}}$.

475. Family of diamond back moth is _________ **Plutellidae**.

476. Which is a highly fecond harmful insect species. **DBM**.

477. DBM is a specific insect of __________ **Cabbage**.

478. IPM is also known as **IPC**.

479. Scientific name of boll weevil is **Anthonomus grandis**.

480. Bt cotton is resistant against which crop **Lepidoptera**.

481. First pesticide used in India was **DDT**.

482. Insecticide which is also known as Gamma-hexachloricyclohexane **Lindane**.
483. ________ is an organophosphate **Diazinon**.
484. Pollyphagous insect-pest is **Locust**.
485. Monophagous insect-pest is? **Rice yellow stem borer, Cotton jassid and Rice weevil**.
486. Vector of rice tungro virus is ______ **GLH**.
487. Vector of Grassy Stunt Disease is **BLH**.
488. Ear cockle of wheat is caused by **Nematode**.
489. Use of pesticide is highest in _________ **Cotton**.
490. Family of rice gandhi bug is **Alydidae**.
491. Management practice of tundu disease includes **Hot water treatment**.
492. Moths are _________ **Usually night fliers**.
493. Butterflies are _________ **Usually day fliers**.
494. Hopper burn (*A. biguttula*) is a symptom of. **Cotton**.
495. Which is potential pest? **Army worm on wheat**.
496. Pest population studies are of _________ **2 types**.
497. Pest population estimates are of _________ **3 types**.
498. Relative estimate includes _________ **Use of traps**.
499. Gundhi bug attacks on **Milking stage**.
500. Biological control of woody aphid is done by applying _________ **Trichograma**.

5

Horticulture

1. Leading cut flower exporter in the world is **Netherlands**.
2. Concept of lawn was developed in **England**.
3. In which garden, arrangement of rocks is main feature **Japanese garden**.
4. Shrubs or trees planted at regular intervals on boundry for fencing **Hedges**.
5. The green carpet for the landscape maintained by growing and mowing grasses. **Lawn**.
6. Thorny fencing plant used as a hedge **Inga dulcus**.
7. Planting of low growing plants along with paths, roads, flower beds, lawns etc. for demarcation and beautification **Edge**.
8. The art of developing the plant or training the plant into different forms or shapes like animals, birds, arches, etc. **Topiary**.
9. The arrangement of colourful potted plants in different tiers around a central object which may be tree trunk, lamp post or pillar **Trophy**.
10. Growing of shrubs in a group **Shrubbery**.
11. A group of ornamental plants used to grow over walls, trellis, arches, pergolas, arbours, pillars, bowers etc **Climbers and Creepers**.
12. An art of growing and training of a plant to a miniature form having a natural look of old age **Bonsai**.
13. The father of rose breeding **Dr. Bhattachaterji**.
14. The fruit of rose is known as **Hip**.
15. Rose can be cultivated up to **5 years**.
16. Commercial propagation method of rose **T' budding**.
17. Floribundas is a cross of **Hybrid tea × Dwarf polyantha**.
18. Yellow coloured rose species is ***R. foitida***.
19. Thornless variety of rose **Chitra**.

20. The single borne rose species **Hybrid tea**.
21. Crimson Glory and Super Star are which type of rose **Hybrid tea**.
22. Rose species having large flowers in clusters **Floribundas**.
23. The growth habit of Chrysanthemum **Perennial**.
24. Commercial propagation method of Chrysanthemum **Root suckers and Terminal cuttings**.
25. Which chrysanthemum allows single bloom on a branch? **Standard Chrysanthemum**.
26. Off-season variety of chrysanthemum **Haldi ghati, Himansu, Jaya and Jwala**.
27. Per hectare of seed rate of marigold **1.2-1.5 kg/ha**.
28. Commercial propagation method of gladiolus **Corms**.
29. Annual carnation is propagated by **Seeds**.
30. True marigold is also known as Calendulas.
31. Golden Age and Crown of Gold are the variety of **African Marigold**.
32. Tuber is commercially propagated by **Tubers**.
33. Queen Elizabeth is a variety of ***Floribundas rose***.
34. The botanical name of hollyhock ***Althaea rosea***.
35. The flower colour of Chrysanthemum **Yellow and white**.
36. Edible part of cauliflower - **Curd**.
37. Blanching is an important process of - **Cauliflower**.
38. Whiptail of cauliflower due to - **Mo deficiency**.
39. Pusa ruby is a variety of - **Tomato**.
40. Swaran roopa is the early variety of **Litchi**.
41. Pajaro: is the variety of - **Strawberry**.
42. Bolero is the variety of - **Marigold**.
43. Richest source of vitamin C is - **Barbados cherry**.
44. Largest importer of cut flower in the world - **Germany**.
45. The dwarf variety of mango - **Amrapali**.
46. California Papershell is the variety of - **Almond**.
47. Fruit of rose known as - **Hips**.
48. Fruit of okra is - **Capsule**.

49. Coconut fat rich source of - **Lauric acid**.
50. Coconut is propagated by - **Seeds**.
51. Mango is mostly propagated through - **Veneer grafting**.
52. Black heart of potato due to - **O_2 deficiency**.
53. Pusa snowball is a variety of - **Cauliflower**.
54. Development of fruits without fertilization is called - **Parthenocarpy**.
55. Development of embryo without fertilization is known as - **Apomixis**.
56. Oleoresin is an important product of - **Chilli seeds**.
57. Kesar (saffron) belong to the family of - **Iridiceaev**
58. Concentration of sugar is used for preservation is - **60-70%**.
59. Mango variety suitable for high density planting is - **Amrapali**.
60. Most salt tolerant fruit crop is - **Date palm**.
61. Fruit repiring hormone is - **Ethylene**.
62. Origin place of rose - **India**.
63. Rose is know as - **Queen of flower**.
64. Rose is Symbol of - **beauty**.
65. Rose is national flower of - **England**.
66. Best time of planting rose is - **September to October**.
67. Commercial method of propagation - **T-budding**.
68. Best time of budding - **November to February**.
69. Fruit of rose - **Rose hips**
70. Dr.BP .PAL is related to - **Rose breeding**.
71. Gulkand is prepared by mixing Petals and sugar in - **1:1** ratio.
72. White flower variety of rose is - **Tushar**.
73. Red flower variety of rose is - **Crimson Glory**.
74. Thorn less variety of rose - **Pusa Mohit, Suchitra**.
75. Increase shelf life of flower **Pulsing treatment** is done.
76. Scientific name of Gladiolus - **Gladiolus hybridus**.
77. Family of Gladiolus **Iridacea**.
78. Propagation method of **Gladiolus Corn / Tissue culture**.
79. Scientific name of the Chrysanthemum - **Dendranthema gradifora**.

80. Family of Chrysanthemum - **Compositae/Asteraceae**.
81. Common name of Chrysanthemum - **Guldaudi**.
82. Origin of Chrysanthemum- **China**.
83. Propagation method of Chrysanthemum is - **Root sucker**.
84. Scientific name of marigold - **Tagetes Spp**.
85. Seed rate of mari gold **One kg/ha**.
86. Fruit can be processed at a temperature of **100°C**.
87. Vegetable can be processed at a temperature of **115-121°C**.
88. Asepsis prevent entry of **Microbes**.
89. Raisins known as **Kismis**.
90. Oxidation can checked by **Antoxidants**.
91. Scientific name of coconut - **Cocus nucifera**.
92. Family of coconut is **Aracaceae**.
93. India rank **3rd** in Coconut production.
94. Kerala share in coconut production is **42%**.
95. Scientific name of Rhinoceros beetles - **Oryctus rhinoceros**.
96. Fan like appearance of leaves in - **Rhinoceros beetles**.
97. Scientific name of Grape - **Vitis vinifera**.
98. Family of grape- **Vitaceae**.
99. Grape is a - **Deciduous crop**.
100. Grape propagated by - **Hard wood cutting**.
101. Element that is universal deficient in grape is - **Mg**.
102. **Tartaric acid** is present in Grape fruits.
103. Arka Trishna & Soma variety of grape is use to - **Wine purpose**.
104. Arka Hans variety of grape is use to - **White wine**.
105. **Arka Kanchan** variety of grape is Late maturity variety.
106. Perlette variety of grape is Mostly grown in **Rajasthan**.
107. **Downy mildew** is major disease of Grape.
108. Scientific name of Aonla - **Emblica officinalis**.
109. Family of Aonla is - **Euphorbiaceae**.
110. Aonla is also known as - **Indian gooseberry**.

111. Propagation by Aonla is - **Patch budding**.
112. Processing vegetable **Tomato**.
113. The new botanical name of tomato **Solanum lycopersicon**.
114. A pigment responsible for red colour in tomato **Lycopene**.
115. Pusa Rubi is a cross of **Sioux × Improved Maruti**.
116. The best combiner variety of tomato **Pusa Rubi**.
117. Pusa Rubi, Arka sourav, Pant bahar and Best of all are **Indeterminate varieties**.
118. The best suited variety for drought condition **Arka Vikas**.
119. Most serious pest for tomato plant **Root knot nematode**.
120. Nematode and Bacterial wilt resistant variety of tomato **Arka vardan**.
121. Most important nutrients required for tomato cultivation **Boron and zinc**.
122. The recommended seed rate of tomato per hectare **300-350 gm/ha**.
123. Harvesting stage of tomato for distant market/transportation **Mature green stage**.
124. Best method of extraction of tomato seed **Alkali method**.
125. Blossom end rot of tomato is due to **Ca deficiency**.
126. Fruit cracking of tomato is due to **B deficiency**.
127. Brinjal is also known as **Egg plant**.
128. The botanical name of acid lime **Citrus aurentifolia**.
129. Break fast fruit is **Grape fruit (*C. paradise* leaf)**.
130. The botanical name of Sweet Orange **Citrus sinensis**.
131. Which citrus species is known as Fancy fruit? ***C. reticulate***.
132. Thornless species of citrus **Tahiti lime (*C. latifolia*)**.
133. Monoembryonic species of citrus **Pumelo**.
134. Polyembryonic species of citrus **Acid lime**.
135. Which species of citrus produces seedless fruits? **Tahiti lime**.
136. The normal planting space of citrus **5-6m × 5-6m**.
137. Commercial propagation method **Seed and Budding**.
138. Rootstock of mandarin orange is **Rangpur lime**.
139. Best method for irrigation of citrus **Ring method**.

140. Kinnow' is a cross between **King × Willow leaf**.
141. Seedless variety of mandarin orange **Satsuma**.
142. Lucknow seedless is a variety of **Lemon**.
143. The banana is commonly known as **Adam's fig and Tree of paradise**.
144. The normal planting space of banana **1.8-2m × 1.8-2m**.
145. Commercial propagation method **Sword Suckers**.
146. Banana inflorescence is known as **Spadix**.
147. Best variety for chips making **Narendran**.
148. Gold finger is a **Hybrid banana**.
149. How much sugar contains in ripe banana? **26%**.
150. Degreening of banana is done by **Ethylene**.
151. Removal of male bud after completion of female phase is known as **Denavelling**.
152. Removal of undesired suckers, done once in 45 days of planting **Desuckering**.
153. Tetrazolium test is used for detection of **Bunchy top virus**.
154. Which crop is commonly known as Single seeded nut? **Litchi**.
155. Litchi is commercial propagated by **Air layering and Seed**.
156. Red pigment in litchi is due to **Anthocyanin**.
157. Sapota is grown in **Tropical climate**.
158. Commercial propagation method of sapota **Inarching**.
159. Most popular varieties of sapota **Kali patti, Pili patti, Oval, Cricket ball, Chatri, Barahmasi**.
160. CO-1 is a cross of **Cricket ball × oval**.
161. Botanical name of Monkey jack **Autocarpus heterophylus**.
162. Jackfruit is commercially propagated by **Air layering**.
163. The popular varieties of jackfruit **Champa, Rudrakshi and Singapore**.
164. Aonla is commercially propagated by **narching**.
165. The popular varieties of Aonla **Banarasi, Krishna, Chakaiya, Hathi jhul, Kanchan, NA-7, 9**.
166. Most widely used training system of aonla **Modified central leader system**.

167. Ber is commonly known as **Poor man's fruit and king of arid fruits**
168. The commercially cultivated variety of ber **Umran**.
169. Early variety of ber **Seb**.
170. Best time for prunning of ber **End of May to Mid June**.
171. Ber fruits are matured at **5-6 months after flowering**.
172. Richest source of Vit-B2 **Bael**.
173. Active ingredient present in Bael **Marmelosin**.
174. The most ideal stage of bael for making preserve **Mature green stage**.
175. Paras is a variety of **Jamun**.
176. Seedless variety of Jamun is **Narendra Jamun-6**.
177. The commercial propagation method of pineapple **Suckers and slips**.
178. Singapore, Mauritious and Giant Kew are the varieties of **Pineapple**.
179. The enzyme contains in pineapple fruit **Bromelin**.
180. The chemical used for inducing flowering in pineapple **Ethrel and NAA**.
181. Queen of temperate fruit **Apple**.
182. Apple bowl of India **Himachal Pradesh**.
183. Apple is commercially propagated by **Tongue grafting and Whip budding**.
184. Redness in apple is due to **Anthocyanin**.
185. Delicious, Rome beauty and Parlin's Beauty are **Late maturing apple**.
186. Diploid variety of apples are **Self fertile**.
187. Usually apple is graded on **6 size**.
188. Discolouration of apple after cutting is due to **Enzymes**.
189. Commercial propagating method of grapevine **Hard wood cutting**.
190. Arka Hans is a cross of **Banglore Blue × Anab-e-shahi**.
191. Most widely accepted training system of grapevine in India **Bower system**.
192. Which is used for improving fruit quality of grape? **20 ppm GA**.
193. Which fruit is commonly known as "Kalpavriksha **Coconut**.
194. Cashew nut and almond are the richest source of **Fat**.
195. Richest source of Iron **Karonda**.

196. Orgin of mango is - **Indo-burma**.
197. **India** is first rank in production of mango (about 50% of world mango produced in india.
198. Mango is extremely rich in **vitamin A**.
199. Mesocarp is edible part of **mango**. (between outer layer and seed).
200. **Sai sugandha** is a variety which is regular bearing also free from spongy tissue disease.
201. Mango plant well grown in well drained **loamy soil**.
202. In India **Kanyakumari and Madurai district** produce 2 time fruits in a year.
203. Mango hopper is one of the major pest of mango.
204. Veneer grafting method of propagation adopted commercially best suited during **monsoon season**.
205. Epicotyl stone grafting is a another latest discovered method of propagation in **mango plant**.
206. Dashahri is one of the best medium ripening variety of mango which grown in north India Majority of **Uttar Pradesh**.
207. Spongy tissue in mango mainly due to **heat convection/higher temperature**.
208. **Alphanso** is most susceptible of spongy tissue.
209. Deficiency of **potassium** cause leaf scorching in the plant leaves.
210. Alternate bearing problem often seen in **Chausa, Dashahri, and Langra variety**.
211. Tolerant to irregular bearing - **Amrapali, Ratna, Neelum etc**.
212. Spongy tissue tolerant variety - **Arka puneet, Arka arun, Arka anmol another** north indian varieties are often do not affected by spongy tissue
213. Boron deficiency causes - **internal necrosis**
214. **Boron deficiency** also caused Black tip in mango.
215. **Black tips don't due to only** Boron, Co, CO2, SO2, Acetylenes gases are also responsible for black tip.
216. Alkathene banding are one of the best method to protect plant from **mango mealy bug**.
217. Olericulture' - is a **'Latin'** word.

218. 'SUN-GROW-496' is a hybrid of **tomato**.

219. 'PBOG-1' A long fruit bottlegourd (Lauki) hybrid, developed from **Pantnagar**.

220. Radish, Tomato and Chilli contain **Vit. 'C'** in enough quantity.

221. Cauliflower (a Latin word) means - Caulis = Cabbage + Floris = flower; the cauliflower is a **Cole crop**. The cole crop comprising to cauliflower, cabbage and Knol-Khol.

222. In cauliflower, buttoning is due to the **deficiency of nitrogen**.

223. The pungency in Chilli is due the presence of active constituent **'Capsaicin'** alkaloid (C18H2O3N with 64·5-65°F).

224. The red colour in tomato is due to **'Lycopin'** pigment, formed at 20-25°C temperature.

225. Sweet-potato is a **short day plant**.

226. **Male sterility** is found in Onion, Carrot and Tomato.

227. Onion, Garlic, Arvi (Colocasia) and Yam are **monocolyledon Vegetables**.

228. For breaking dormancy of Potato tubers, **thiorea** is used.

229. Tristeza Virus is spread due to **aphids**.

230. 'Pusa Jwala' Chilli is a cross of **NP-46-A × Pusa Red**.

231. For commercial crop production of tomato, the growth regulators viz. **NAA** (30 ppm) and **GA** (20 ppm) are used as spray, resulting there by higher crop yield.

232. 'Ricyness' in cauliflower develops due to application of excess **nitrogen** and higher **relative humidity**.

233. The **yellow colour** in onion is developed through CCrr genotype.

234. 'Cardamom' is said to be the **'Queen of Spices'**.

235. Pumpkin Day celebrated on-**29 Sept. in Germany**, 320 kg weight of a pumpkin/Kaddu was recorded highest in world and in England 'Kheera' say - Cucumber length was 43 inches largest in world.

236. 'Hydroponics' is a '**Greek**' word and related to without land cultivation i.e. crop cultivation in water e.g. waternut (Singhada).

237. The seeds of synthetic varieties are changed after every **5-6 years**.

238. The seeds of hybrids (varieties) of crops are purchased by the farmers **every year**.

239. The crop of tomato has the quality of both **thermo and photo-sensitive**.

240. The dehydration ratio in potato is **7:1**.

241. The dry ratio in Okra (Lady finger) is **12:1**.

242. Sugarbeet is **hexaploid**.

243. 'Whip-tail' disease in cauliflower is caused due to the deficiency of **molybdenum (Mo)**.

244. 'Browning' disease in cauliflower is caused due to the deficiency of **boron (Bo)**.

245. Pungency in onion is due to the presence of **Allyle Propyl di-sulphide**.

246. 'Mosaic' in potato is caused due to **virus**.

247. Seed treatment through **'Bavistin'** controls effectively the seed borne diseases in crops.

248. The word 'curd' is related to cauliflower, wherein the colour of head becomes **curdy white**.

249. 'Azad Nutan' is a variety of **Bottle-gourd**.

250. 100 g fresh carrot contains approx **3150 (IU) Vit 'A'**.

251. **Vit. 'A'** is abundently found in green leafy vegetables.

252. All vegetables have generally **more or less** all vitamins.

253. Night-blindness in human is caused due to deficiency of **Vit. 'A'**.

254. Onion (green/bulbs) is the best source of **Vit. 'B'**.

255. The best sources of **'Fe' (iron)** are-carrot, bittergourd, onion and tomato.

256. Nearly **285 g** vegetables per day per capita in human diet for body is recommended.

257. Cauliflower is sensitive for **excess of acidity of soil**.

258. Cauliflower requires more **boron (Bo) and molybdenum (Mo)**.

259. Cabbage requires more plant nutrients (i.e. heavy feeder) particularly **the N and K** nutrients in excess amount.

260. The seed rate of Knolknol (Ganthgobhi) in nursery is maintained about **1250 g** for one hectare transplanting.

261. Ripened tomato contains about **94%** water, which is more suitable for good source of **Vit. 'C'; Vit 'A' and Vit. 'B'**.

262. Cucumber is the oldest cultivated vegetable crop; which is possibly originated from **India**.

263. The weevil (Cylas formicarius) is most serious insect of **Sweet potato**.

264. The scientific name of Mango hopper ***Amritodus atkinsoni***.
265. The scientific name of Mango mealy bug ***Drosicha mangiferae***.
266. Sticky bands around tree trunks provide protection against **Mango mealy bug**.
267. The scientific name of Banana Stem weevil ***Odoiporus longicollis***.
268. The scientific name of Guava fruitfly ***Bactrocera diversus***.
269. The scientific name of Fruit sucking moth of citrus ***Otheris materna***.
270. The scientific name of Lemon butterfly ***Papillio demoleus***.
271. The scientific name of Citrus Psylla ***Diaphorina citri***.
272. The scientific name of Papaya fruitfly ***Bactrocera/Dacus dorsalis***.
273. The scientific name of Fruit Borer of pomegranate ***Conogethes punctiferalis***.
274. The scientific name of Woolf aphis of apple ***Eriosoma lanigerum***.
275. The scientific name of Tomato fruit borer ***Helicoverpa armigera***.
276. The scientific name of Brinjal fruit and shoot borer ***Leucinodes orbonalis***.
277. The scientific name of fruit and shoot borer of Okra ***Earias vitella***.
278. The scientific name of Chilly thrips ***Thrips tabaci***.
279. The scientific name of Red Pumpkin beetles of cucurbits ***Raphidopalpa foveicollis***.
280. The scientific name of Cucurbits fruitfly ***Dacus cucurbitae***.
281. The scientific name of Diamond Back Moth (DBM) of cabbage ***Plutella xylostella***.
282. The scientific name of Cabbage head borer ***Hellula undalis***.
283. The scientific name of Khapra beetle or Wheat beetle ***Trogoderma granarium***.
284. The scientific name of Red flour beetle ***Tribolium castaneum***.
285. The scientific name of Pulse beetle ***Callosobruchus chinensis***.
286. The scientific name of Rice moth ***Corcyra cephalonica***.
287. 1st Sequenced vegetable crop - **Cucumber**.
288. 1st Gynoecious line of Cucumber is - M.SU. 713-5 (Shogoin X Wisconsin SMR 18) developed by **C.E. Peterson (1960)**.
289. 1st Gynoecious Hybrid in cucumber - **Pusa Sanyog in 1971**.
290. 1st Tropical Cucumber Hybrid - Phule Prachi by **MPKV, Rahuri**.

291. 1st Extra Early Parthenocarpic variety of cucumber - **Pusa Seedless**.

292. 1st True Mini Watermelon Cultivar - New Hampshire Midget Developed by **A.F. Yeager**.

293. 1st Fusarium Wilt Resistant Watermelon - Conquerer Developed by **W.A. Orton**.

294. 1st Horticultural Classification of Melons - Naudin (1859); Modified by **Munger and Robinson (1991)**.

295. 1st F1 Hybrid of Muskmelon- Punjab hybrid (MS-1 X Hara Madhu) in 1984 - **Resistant to PM**.

296. 1st Muskmelon Variety Grown under Net-house in N.Indian Plains - **Pusa Sarda**.

297. 1st Early Maturing variety of Round melon - **Pusa Raunak**.

298. 1st Early Maturing variety of Long melon - **Pusa Utkarsh**.

299. 1st Hybrid of Pointed Gourd - **CHES Hybrid-1** (Resistant to Fruit Fly).

300. 1st Improved Variety of Ash Gourd - **Pusa Ujjwal** (Petha Preparation).

301. 1st Inter-varietal Hybrid of Greater Yam - **Shree Shilpa**.

302. 1st Coleus Variety Grown in Kerala - **Shree Dhara**.

303. 1st Report of Enation Leaf Curl Virus(Vector : White Fly) **- IIHR**.

304. 1st Machine-Harvestable Tomato cultivar was developed by **G.C. Hanna**.

305. 1st Male sterility in tomato was reported by **C.M. Rick**.

306. 1st Tomato Resistant Cv. **Pan America**.

307. 1st Triple disease resistant variety of Tomato **- Arka Rakshak**.

308. 1st Hybrid of Tomato available cultivation in **1973**.

309. 1st Root-Knot Nematode Resistant variety of Tomato: **Sel-120**.

310. 1st Report of Heterosis in Brinjal : **Kakisaki (1931)**.

311. 1st Male Sterility in Brinjal : **Jasmin (1954**).

312. 1st Functional Male Sterile Line in Brinjal: **UGA1-MS** (Cv. Florida High Bush), **Pathak and Jaworski (1989)**.

313. 1st Reported Brown Anther CMS in Cv. "Tender Sweet" by **Welch and Grimbell in 1947**.

314. 1st Carrot fly Resistant Variety - **Flyaway in 1993**.

315. 1st Black Coloured Variety of Carrot India - **Pusa Asita**.

316. 1st Temperate Hybrid Developed Using CMS - **Pusa Nayanjyoti-1**.

317. 1st Public Sector Hybrid Using CMS is - **Pusa Vasudha**.
318. 1st Genetically Improved Variety of EFY in the world is **Sree Athira**.
319. 1st Genetically Improved Variety of Taro - **Sree Kiran**.
320. 1st Reported SI in Cole Crops by **Bateman (1954)**.
321. 1st Cabbage Hybrid - **Nagoka in Japan (1951)**.
322. 1st CMS based F1 Hybrid in Cauliflower - **KTH-27**.
323. 1st Reported Male Sterility in Radish - **Ogura 1968 in Japanese Radish Type**.
324. 1st Purple Fleshed Radish Variety is **Pusa Sagarika**.
325. 1st Hybrid of Knol-khol - **Roggli**.
326. 1st Reported Little Leaf of Brinjal - **Thomas and Krishnaswamy**.
327. 1st Discovered CMS in Chilli by Peterson in **PI164835**.
328. 1st President of Central Potato Research Institute - **Dr. S.Ramanujan**.
329. 1st International Flower Auction Centre was launched in **Bangalore**.
330. 1st Rank Foliage Plant in Global Market - **Dieffenbachia**.
331. 1st Rank Cut Flower in Global Market - **Rose**.
332. 1st Rank Pot Plant in Global Market - **Hedera**.
333. 1st Rank Dry Flower in Global Market - **Helichrysum**.
334. 1st Rank Cut Green in Global Market - **Asparagus**.
335. 1st Hybrid in Ornamentals : "**Primma Donna**" in Begonia semperflorens in 1909 by Benery Seed Company, Germany.
336. 1st Evidence of Ornamental Tree, **The pipal** (Ficus rugosa) comes from Mohan-Jo-Daro.
337. 1st World (world's 1st) Vertica Forest - **Bosco Verticale in Milan**.
338. 1st Islamic garden in India was Developed by **Feros shah**.
339. 1st Mughal garden was developed by Baber at Agra Called "**Ram Bagh**".
340. 1st Tomb garden made in India - **Humayun's Tomb**.
341. 1st Garden made by Akbar - **Fatehpur Sikri (Agra); 2nd at Sikandra (Agra)**.
342. 1st Botanical garden in Tropics was **at Pamplemousses at Mauritius (1735); 2nd Kolkata botanical garden (1787) (Note: Both gardens are Informal Type (i.e) English garden style)**.

343. 1st botanical Garden in world was at **Padua, Venice, Italy in 1543**.
344. 1st Professional floral designer's - **Greek**.
345. 1st Bio-Aesthetically planned City - **Chandigarh**.
346. 1st Reference of Rose Perfume - **"Charaka Samhita"**.
347. 1st Bonsai Garden - **Kamala Nehru Park**.
348. 1st Colchicine Induced Tetraploid synthetic species of Day lily is **Hamerocallis washingtonia**.
349. 1st Imported Anthuriums are **Pastel Coloured**.
350. 1st widely cultivated Anthurium Pot Plant - **Anthurium scherzerianum**.
351. 1st Chrysanthemum known to USA - **"Dark Purple"** Imported by a Scientist John Stevens in 1798.
352. 1st African Marigold Variety - **Climax and Toreador**.
353. 1st Introduction of Dahlia in 1857 by Agri-Horticultural Society of **India, Kolkata**.
354. 1st Gerbera was discovered by **R. Jemson in South Africa in the year 1878**.
355. 1st Flower grown in space - **Zinnia**.
356. 1st cultivar of Carnation : **Souvenir La Malmaison in 1857**.
357. 1st Indian company to grow carnations in International standards - **FEROZE and also to be registered at VBA**.
358. 1st GM floriculture crop having Violet Blue colour variety - **Moonlight (carnation)**.
359. 1st person to grow gladiolus in India - **Mr. Charles Gray**.
360. 1st fragrant variety of gladiolus - **Lucky Star** (2n=40; Bigeneric).
361. 1st smoky variety of gladiolus - **Prince of India**.
362. 1st chemical used to break corm dormancy - **Ethylene** Chlorohydrin by Denny (1938).
363. 1st *in vitro* propagation of gladiolus - **Ziv *et al.*** (1970).
364. 1st species of Bougainvillea introduced in India - ***B. spectabilis***.
365. 1st Bougainvillea variety by S. Persy Lancaster in India - **Scarlet Queen** (1920).
366. 1st Orchid Bletia purpurea (1731) was grown by **Peter Collinson**.
367. 1st Artificial hybrid of Orchid Calanthe Domini (*C. mausacata* x *C. frucata*) by **Dr. J. Dominy (1852)**.

368. 1st Ornamental Orchid - **Cypripedium reginae**.

369. 1st Bigeneric hybrid of Orchid - **Phaius Grandiflora X Calanthe vesita**.

370. 1st Sexageneric hybrid of Orchid - **'Brilliendeara Gary'** registered in 1982.

371. 1st Meristem tip culture in Orchid is Cymbidium sp. **by Morel (1960)**.

372. 1st Tissue culture in Orchids was done by **'Morel'**.

373. 1st Tissue culture laboratory for Orchids by **A.V. Thomas in 1984**.

374. 1st hybrid of Snapdragon(Antirrhinum majus) - **Christmus Cheer**.

375. 1st consignment of Rose was sent by IARI to Paris, **Rotterdam, and Furkfurt in 1959**.

376. 1st World's (World's 1st) blue rose - **Applause**.

377. 1st HT Rose - **La France by Gulliot (1867)**.

378. 1st HT Yellow Rose - **Soleild**.

379. 1st Hybrid Perpetual - **Princess Helene**.

380. 1st Floribunda - **Rodhatte by Poulsen (1912)**.

381. 1st Grandiflora - **Buccaneer (1952)**.

382. 1st Polyantha - **La paquerette (1875)**.

383. 1st Rose variety by **Dr. B.P. Pal is Sherbet**.

384. 1st Indian Rose breeder who raised Cv. **Dr.S.D.Mukherjee in 1935 by B.K. Roy Choudhary**.

385. 1st Climate controlled greenhouses established in India by **Indo-American Hybrid Seed Company in 1965**.

386. 1st Chief scientific officer of UPASI(valparai) - **Dr. W.S. Shaw**.

387. The edible part of potato is - **Tuber**.

388. Vegetable production for local consumption is: **Market gardening**.

389. Foliage plants with extremely tender and watery stem are - **Succulents**.

390. Most of the summer vegetables are - **Annual**.

391. Summer vegetables can be grown out of season in a climatic region:**Lower Sindh**.

392. In seed formation the inner integument remains thin and develops into: **Tegmen**.

393. During seed formation, a sear left by the funiculus is termed as. **Operculum**.

394. Endosperm like tissue (2n) in seed developing from nucellus is **Perisperm**.
395. Reduction or elimination of dormancy in crop plant is due to - **Acclimatization**.
396. Dormancy due to factors within the embryo is - **Endogenous**.
397. Secondary dormancy in non dormant seeds is due to - **Temperature**.
398. Stage at which seed achieves it maximum dry weight and has maximum germination potential and vigor is - **Physiological maturity**.
399. Tetrazolium test to check seed viability is **Chemical test & Germination test**.
400. Researchers describe seed moisture contents in terms of - **Dry weight & Wet weight**.
401. Celery and lettuce at high temperature become **Thermodormant**.
402. Excessive soil moisture, late in the season cause. **Decrease in TSS increase in dry** matter.
403. Most of following category crops are shallow rooted **Summer**.
404. Deep rooted vegetables require - **Less frequent irrigation and More water at each irrigation**.
405. Fruit of following vegetables is edible - **Summer**.
406. Mulches are mainly used to reduce - **Transpiration, Evaporation and Evapotran spiration**.
407. What type of mulches reduces soil temperature? **Organic mulch**.
408. For hybrid seed production flowers are completely suppressed on female plants **Staminate**.
409. Recommended cucumber cultivars under tunnel production technology are: **Gynoecious**.
410. Flowering in water melon is unaffected by light duration as it is **- Long day short night plant**.
411. Nature of pollination in cucurbits is **- Highly cross pollinated**.
412. Ginger is commonly cultivated as - **Annual**.
413. Cross pollination is essential in following type of vegetables. **Dioecious**.
414. Cross pollination in plants bearing bisexual flowers is due to **Dichogamy, Herkogamy & Heterostyly**.
415. In anemophilous plants, pollens are disseminated by means of - **Insects**.
416. In pollen development Gametocides induces a problem of - **Sterility**.
417. Production of seeds without fertilization is termed as - **Parthenogenesis**.

418. Most of the hybrid watermelon and cucumber are - **Colchiploids**.
419. Removal of off type plants from a seed field is termed as - **Rouging**.
420. The most important seed component affecting seed /viability during storage is - **Temperature**.
421. Potato is generally propagated through **Seed potato, Potato seed and True Potato Seed**.
422. Isolation during seed production is done by - **Plantation at a distance**.
423. The most common pollinator insect in Pakistan is **Honey bees**.
424. Seed production in biennial vegetables is mostly done by **- Seed to seed method, *in situ* method & Replanting method**.
425. Male sterily is not due to - **Nutritional factor**.
426. Hybrid vigor in vegetables is manifested by **High yield, Uniformity in size and Uniformity in maturity**.
427. Water melon for seed should be harvested when **Tendrils wither**.
428. In Pakistan, vegetables for seed production are harvested by **Manual method**.
429. During storage of vegetable seeds, seed moisture should be **Less than 10%**.
430. The poorest seed storage capacity is in **Onion**.
431. Nitrates are absorbed at pH below - **5**.
432. Hormones which control apical dominance are - **Cytokinins**.
433. Tetraploid watermelon must be pollinated by plants to produce a seedless watermelon **Diploid**.
434. Which vegetable cannot be grown as an off season vegetable - **Potato**.
435. Lower temperature than normal convert male flowers into perfect flowers in **Papaya**.
436. Tomato is a crop. **Self pollinated**.
437. Turmeric is propagated by - **Rhizome**.
438. Sweet potato is propagated by - **Stem cuttings & Tubers**.
439. Self pollination is a form of - **Inbreeding**.
440. Cross pollination is associated with - **Chasmogamy**.
441. Completion of pollination and fertilization in unopened flower bud is termed as **Cleistogamy**.
442. Maturation of anthers and stigma at the same time refers to **- Homogamy**.

443. Fertilization after opining of flower refers to - **Chasmogamy**.
444. Male sterility promotes - **Chasmogamy**.
445. Exposure of seed to low temperature before germination is termed as - **Stratification**.
446. Imbibed or germinated seeds subjected to cold temperature is termed as - **Vernalization**.
447. Which of following vegetable does not require specific low temperature before flowering - **Brinjal**.
448. Edible part of cauliflower is - **Vegetative bud**.
449. Edible portion of cabbage is - **Leaves**.
450. Cole crops are - **Hardy**.
451. Which of the following crop has lowest water contents - **Cauliflower**.
452. The commercial variety of watermelon is - **Sugar baby**.
453. Vine of crop of bottle gourd is not staked during - **Summer season**.
454. Which one is not the best source of protein? **Potatoes**.
455. The cause of hotness in Chillies is - **Capsaicin**.
456. Which crop is most susceptible to viral infection - **Tomato**.
457. Which of the following vegetable has a dioecious behaviour - **Bitter gourd**.
458. Consecutive use of garlic will reduce the level of - **Cholesterol**.
459. Bulb production in onion is dependent upon - **Day length**.
460. Red colour of carrot is due to - **Carotene**.
461. Colour of carrot roots is affected by - **Soil moisture**.
462. Radish belongs to family - **Umbelliferae**.
463. For an early harvest, cucurbits are grown on - **Sandy soils**.
464. Tomato fruit is botanically a - **Berry**.
465. The portion of axis below the cotyledons in seed is - **Hypocotyl**.
466. Which type of flowers can not produce seeds and/or fruit? **Staminate**.
467. Degeneration of above ground parts in ginger is due to - **Senescence**.
468. In spinach, pollination takes place by means of - **Wind**.
469. Optimum temperature for cool season vegetables is - **18-24°C**.
470. Which one is not an appropriate approach to control frost? **Avoidance**.
471. Tomato, chillies and brinjal are grouped as - **Solanaceous crops**.
472. Color development in tomato fruit is due to - **Lycopene**.

473. Cucurbits consumed at immature stage are grouped under - **Gourds**.
474. Term vine yard is used for - **Grape plantation**.
475. Term viticulture is used for - **Grape cultivation**.
476. Home gardening deals with vegetable production for - **Home consumption**.
477. Truck gardening means vegetable production for - **Distant places**.
478. Tip burn of lettuce occurs at high air temperature when the soil is - **Cool**.
479. Organic crop cultivation is - **Without fertilizers**.
480. Thinning in vegetables is practiced to maintain - **Plant x Plant distance**.
481. Transplanting means - **Shifting of seedling**.
482. Shrubs fall in the category **of 7-15 feet high**.
483. Deciduous plants are mostly grown in - **Hilly area**.
484. Vine mean climbing/trailing plants with - **Herbaceous stem**.
485. Monoecious plants have male and female sexes/flowers **on Same plant**.
486. Hardy crops/plants are acclimatized against - **Frost & Extreme heat**.
487. True fruits developed from - **Ovary**.
488. Drupe fruits have all the three ovarian layers - **Separated**.
489. Creeping vegetables crops are grown on - **Beds**.
490. Mango is a **Drupe fruit**.
491. Citrus is a special type of **Berry**.
492. Dicots could be identified on the basis of - **Reticulate leaf venations**.
493. In binomial nomenclature system, 1st name is - **Genus name**.
494. Most of the summer vegetables belong to family - **Cucurbitaceae**.
495. Cultivar means -**Some cultivated variety**.
496. Pears and beans are rich source of - **Protein**.
497. Potato and sweet potato are well known for -**Carbohydrates**.
498. Leafy vegetables provide -**Iron**.
499. Iron is a carrier of -**Oxygen**.
500. White brinjals are commonly grown for - **Ornamental purpose**.
501. Indian Institute of Arid Horticulture is located at Bikaner.
502. CITH is located at Srinagar.

6

Agriculture Terminology

4H: An agricultural club for children that features hands-on learning

Abiota: The nonliving component of an ecosystem, including the soil, water, and air.

Aborigines: The original or native inhabitants of a country or region.

Absorption Losses: Loss of water from a canal or a reservoir by capillary action and percolation and in case of canal during the process of delivery.

Absorptive capacity: The maximum amount of waste that can be absorbed by the environment.

Acclimation: Adaptation to changing or new conditions.

Acid precipitation: Rain, snow, fog, or dew containing sulfuric and nitric acids produced by fossil fuel combustion.

Acid soil: A soil which is deficient in available bases, particularly Ca and which give an acid reaction when tested by a standard method.

Acre foot water: The amount of water that would cover an area of land to a depth of one foot assuming no seepage evaporation and run off.

Acre inch day: Term used principally in irrigated section of united state for measuring quantity of flow of water. It is equal to a flow which will cover one acre to a depth of one inch in a 24 hours period or 0.042 cubic feet per second.

Acre inch: It is a measure of quantity of flow of water and is equal to the flow which will cover one acre to a depth of one inch.

Acre: (43560 sq. ft) an area of land about 220 feet long and 198 feet wide.

Active solar heating: The use of solar panels to collect and concentrate the sun's energy for heating.

Adaptation: Biological modification that allows species to better exist in a specific environment.

Additives: Chemicals added to food, often con- sidered to represent a threat to human health.

Adiabatic: A condition in which heat is neither gained nor dissipated.

Adobe soil: These soils are formed by the broken material of rocks transported by both wind and water.

Aerosol: A suspension of particles in the atmosphere.

Afforestation: Establishment of forest in an area not previously forested.

Aflatoxins: Carcinogenic toxins produced by molds in stored crops.

Agar: A substance made from seed weed and used in the solid culture.

Age distribution or structure: The proportions of a population falling into three distinct groups: preproductive, reproductive, and postproductive. If the population is divided equally among the three groups, the population growth is stable. When a large proportion of the population is in the preproductive stage, the long-term trend is population growth. If a large proportion of the population is in the reproduc- tive stage, a baby boom is experienced.

Agency: Extent to which humans are able to create and change the world in which they live.

Agenda 21: One of several documents emerging from the Earth Summit in Rio de Janeiro in June 1992. Major issues of environment and development were examined, including poverty, population, and human health.

Agent orange: A mixture of several herbicides that is considered to be a carcinogen because it is contaminated with dioxin. The name comes from the orange-banded barrels in which it was marketed.

Agribusiness: a way of farming that combines agriculture and business and usually involves large amounts of land, animals, and expensive technology be

Agribusiness: Industrialized agriculture con- trolled by corporations.

Agricultural economy: An economic system based primarily on crop production.

Agricultural revolution: A shift that took place 10,000 to 12,000 years ago and was characterized by the movement of human activity from hunting and gathering to agriculture.

Agriculture: It is an art, science and business of raising crops and rearing of animals through exploring the natural resources with the coordination of socio economic infrastructure to meet the basic necessities of life i.e. food, feed, fiber and shelter.

Agroforestry: An agricultural system based on the cultivation of trees with other crops.

Air drainage: The flow of cold air down a hill. Plant crops that are sensitive to late spring and early fall frost on slopes, preferably south-facing, so that cold air will drain downhill and settle in the low spots below the planted area

Air layering: A propagation method. A cut is made three-quartets of the way through the stem at an angle of 45 degrees. A toothpick is inserted to keep the cut from sealing itself closed. The cut is wrapped in sphagnum moss that has been soaked in water and wrung out. This is then wrapped in plastic and tied at the top and bottom with ties.

Air pollution: Contamination of the air with solids, liquids, or gases that may be hazardous to humans or other living organisms. The five primary pollutants are carbon monoxide, hydro- carbons, nitrogen compounds, particulate matter, and sulfur dioxide.

Allelopathy: Phenomenon involving the release of certain chemicals from plant parts into the environment which may when present in sufficient amounts, inhibit or suppress the germination or growth of the plants in the neighborhood. Alluvial soil: These are the soils which are formed by the deposition of broken material of rocks transported and deposited by water of streams and rivers.

Alternate leaf arrangement: Leaves are arranged at one leaf per node on different sides along the stem.

Alternative agriculture: Agriculture based on reduced use of chemical fertilizers and pesti- cides, increased use of crop rotation, and re- duced tillage of the soil.

Alternative crops: Nontraditional crops that can be grown in an area to diversify rotations and increase income.

Alternative energy: Energy produced from sources other than fossil fuels (solar, wind, hydroelectric, geothermal, and biomass).

Altitude: Height from sea level

Altruism: Seeking the good for others.

Amazonia: The Amazon Basin area of South America, consisting of 5-7 million kilometers of grasslands, wetlands, shrublands, lakes, and tropical forests.

Ambient: Surrounding or outside.

Amoebic dysentery: A human disease caused by one-celled parasites called amoebae.

Ancient forests: Forests that have never been cut and typically consist of trees 250 years of age and older.

Animal rights: The belief that animals have rights similar to those afforded to humans.

Annuals: Plants that complete their life cycles in one year or less.

Anther: The part of the flower (on the upper part of the stamen) in which pollen is produced.

Anthracnose: A type of leaf or fruit spot disease caused by a fungus.

Anthropic: Relating to the period during which humans have existed on earth.

Anthropocentric ethic: The belief that only humans have value and that the environment exists solely for the benefit of humans; nature has no rights.

Anthropogenic: Based on human activities; often used to refer to environmental changes caused by human activity.

Application losses: Water losses through percolation or run off.

Appropriate technology: Labor-intensive, small-scale, production methods using renewable energy.

Aquaculture: The farming of fish for human consumption.

Aquifer depletion: Depletion of water of an aquifer resulting from withdrawal that is greater than natural or artificial recharge.

Aquifer: A rock, gravel, or sand formation in which water is collected. An aquifer is not an underground lake, but it very much resembles a soaked sponge.

Arable farming: The term arable farming refers to system in which only crops that require cultivation of the soil are grown.

Arable land: Land that can be cultivated.

Arboriculture: Intensive cultivation of individual trees possibly for fruits gums and resins.

Arid region: The region where total rain fall is less than natural evapo-transpiration rate.

Arid: A condition in which less than 10 inches of rain falls each year and the level of evaporation is greater than the level of precipitation.

Aridity: It is the characteristic of a region where there is low average rain fall or 100% available water. It is permanent feature of region.

Arithmetic growth: An increase in some phenomenon at a constant rate over a specified time period.

Arms: In a grape plant, all main branches two years and older.

Artificial fertilizer: A chemical added to soil to enhance crop production.

Artificial recharge: Adding water to an aquifer.

Asexual reproduction: Duplicating a plant from any cell, tissue or organ of that plant.

Assimilative capacity: The ability of a water body such as a lake or stream to purify itself of pollutants.

Atmosphere: The air that surrounds the earth and is bound by the earth's gravitational attraction.

Atmospheric inversion: A situation in which a layer of warm air traps pollutants. Basin cities such as Denver and Los Angeles experience this problem.

Available water: The water retained in a soil which represents the difference between field capacity and the permanent wilting percentage is called available water.

Average life expectancy: The number of years that an average person can expect to live.

Axillary buds: Buds that are borne laterally on the stem in the axils of the leaves.

Bacteria: Microorganisms that live in the soil and convert nutrients into forms usable by plants.

Balance of nature: An idea popularized by George Perkins Marshall that all life is interrelated and in balance. This idea has been re-

Barani soil: When the source of irrigation to crop is only the rain water that is known as Barani soil.

Base period: Period of time in days from the first watering of crop before sowing and the last watering before harvesting.

Basic needs: The basic items and services needed by an individual to ensure a reasonable standard of living.

Basic seed: Is the progeny of pre-basic seed produced so as to maintain genetic purity and identity.

Basin: Flat area of land surrounded by low ridges or bunds.

Biennials: Plants that require 2 years, or 2 growing seasons with a dormant period in between, to complete their life cycles.

Bioaccumulation: The accumulation of pollutants in an organism; sometimes referred to as bioconcentration.

Biocentric ethic: The idea that nature, not humankind, is the measure of all things.

Biocide: An agent that kills many organisms in the environment.

Biodegradable: Capable of being broken down into basic elements as a result of bacterial or other microbial action.

Biodiversity: The degree of species richness and natural genetic variation.

Biogas: Methane gas produced by animal and human dung, crop residues, and other organic matter; can be used as a fuel or fertilizer.

Biogeochemical cycle: The cycling of chemi- cals or nutrients between abiotic and biotic sectors of the biosphere. Elements involved in biogeochemical cycles include carbon, nitrogen, sulfur, and phosphorus.

Biological amplification: The accumulation of higher levels of pollutants in organisms higher up in the food chain.

Biological control: Reducing pests by utilizing other organisms. For example, controlling gypsy moths by using Bacillus thuringensis, a disease that affects the *gypsy* moth and is readily available in garden centers.

Biological control: The use of natural enemies or diseases to control pests.

Biological diversity: See biodiversity.

Biological evolution: Changes in the gene pool of a species over time.

Biological yield: It is the total dry matter produced by a plant as a result of photosynthesis and nutrient uptake minus that lost by respiration.

Biological yield: It is the total dry matter produced by the crops after synthesis minus respiration losses.

Biomass energy: Energy derived from plant matter.

Biomass: All living matter in an area and stored energy in an organic form like wood.

Biome: A large ecosytem that has distinct climate, geology, and organisms; e.g., desert, tundra, grassland, savanna, woodland, coniferous forest, temperate deciduous forest, and tropical rain forest.

Bioregion: An area defined by natural ecological systems, such as a river watershed.

Bioremediation: Restoring a natural area by the addition of living organisms (e.g., plants or bac- teria).

Biosphere: The natural system of the earth and the atmosphere that supports life.

Biota: The plant and animal life of an ecosystem; often referred to as flora and fauna.

Biotechnology: Biological manipulation of living organisms to produce foods, drugs, and other products for humans.

Birth cohort: A group of people born during a specific time period.

Birth rate: The number of live births in a given year divided by the midyear population.

Bison: A large grazing animal that once reigned over the grasslands of North America but was slaughtered in great numbers during the 1870s.

Blade: The flattened, green portion of the leaf.

Blind hoeing: Hoeing before a crop germinates.

Blocked development: Economic development in less developed countries that is impeded by developed countries.

Bolting: Going to seed. Plants will produce a flower stalk and seeds, and then die before the end of the season. Lettuce and other crops will bolt during very hot temperatures. Unless you are growing your crop for the seeds, you want to delay bolting.

Botanical variety : When a group of plant occurring in nature is different from the general species originally described and the botanical binomial name is not enough to identify it is called as botanical variety.

Botrytis blight: A gray, felt-type mold that covers parts of a plant and causes stunting, dieback, and distorted growth.

Bovine: Cattle

Branch: The part of the tree that arises from the trunk and supports shoots, twigs and leaves. Budding. A type of grafting in which a non-fruitful (vegetative) scion is placed in a stock plant.

Breed: To produce offspring by giving birth or by hatching

Breeder nuclear fission: A nuclear fission pro- cess in which new radioactive fuel is produced. It is considered dangerous because the chain reaction is difficult to control.

Broadcasting: Manual spreading of seed in the field and mixing of the spread seed by ploughing or planking the field.

Brundtland Commission: The World Commis- sion on Environment and Development chaired by Norwegian Prime Minister Gro Harlem Brundtland.

The Commission's report, *Our Common Future* (1987), popularized the notion of sustainable development.

Bubble policy: A policy that allows polluters to discharge more pollutants at one source, if an equivalent reduction occurs at other sources.

Buds: An undeveloped stem, consisting of a tiny bundle of cells, from which leaves, lateral buds, flower parts or all three will arise.

Bulb: A budlike structure that consists of a small stem with closely crowded fleshy or papery leaves or leaf bases.

Bulbil: An aerial bulblet.

Bulblet: Miniature bulbs that develop from meristems in the axils of scaly bulbs.

Bull: A sexually mature male bovine

Bushel: A measurement of capacity or size

C3 plants: Plants which fix CO_2 in three C molecule and do not use temperature and water as efficiently as C_4 plant. e.g wheat, rice, cotton.

C4 plant: Plants which fix CO_2 in to a four C molecules .e.g sugarcane, maize, sorghum.

California condor: A bird once found in the mountains of California but now virtually extinct.

Canal: A waterway built to let boats navigate the waters

Canes: Previous season's growth from the arms or trunk of a grape plant; or, the supporting part of the raspberry plant, similar in function to the trunk of a tree.

Capillary Water: It is that water which is held by surface forces (adhesion, cohesion, surface tension) or films around the particles in angles between them and in capillary pores.

Capillary Water: It is the soil water in excess of hygroscopic water. This exists in the pore space of the soil by surface tension or molecular attraction against gravitational forces. It is only water available for plant growth and development.

Capital goods: Accumulated items used to produce other goods and services.

Capitalism: An economy based on private enterprise and the use of markets for allocating economic resources.

Capitalist economy: An economic system in which the means of production are owned privately.

Capsule: Simple, dry, dehiscent (see definition of dehiscent) fruit with two or more locules (the cavities of the ovary of the pistil of a flower) which split in various ways.

Car emissions: Chemicals produced by the internal combustion engine and considered hazardous. These include carbon monoxide, hydrocarbons, and nitrogen oxides.

Carbon cycle: The process by which carbon, the chemical foundation of living organisms, circulates throughout the natural world. This is only one of several different biogeochemical cycles.

Carbon dioxide: A gas that is an important part of the carbon cycle. Plants absorb CO_2 during photosynthesis, and plants and animals produce it as an end product of respiration. It plays an important role in controlling the earth's surface temperature.

Carbon sink: A part of the biosphere that ab- sorbs more carbon dioxide than it releases; e.g., oceans and rain forests.

Carbon tax: A tax imposed on fossil fuels ac- cording to the amount of carbon contained in them.

Carcinogen: An environmental agent, such as a pesticide, that causes cancer.

Carnivore: An animal or plant that feeds on and digests animals.

Carrying capacity: The total population that a particular area can support at a subsistence level.

Cash crop: Agricultural produce marketed for cash rather than retained for household use.

Cash crop: What a farmer raises, crop or livestock, to sell for money

Catchment's Area: The area which drains the rain water falling on it, via streams and rivers, eventually to the sea or into a lake.

Catkin: A type of spike arrangement of flowers on the floral axis that has unisexual flowers made up of sepals and petals.

Cellulose: Component of plant cell walls that is not digestible by most animals

Centrally planned economy: An economy whose investment and production are coordinated by a central government body.

Cereal Crops: A cereal is defined as crop grown for its edible seed. These crops are also known as grain crops e.g. wheat, Rice, Maize etc.

Certified Seed: It is the progeny of basic seed and is produced by registered growers of seed producing agencies.

Certified seed: The progeny of registered seed stock. It is the final stage in the expansion program and is certified with a metal seal and blue tag.

CFCs: See chlorofluorocarbons.

Chemical: An element or compound naturally occurring or created by humans.

Chernobyl: A nuclear power plant in the former Soviet Union that suffered a serious accident in 1986. Estimates are that 5,000 to 150,000 people who lived in the area will die prematurely.

Chilling injury: Damage to certain horticultural products, such as banana, papaya, cucumber and sweet potato, which results from exposure to cold but above-freezing temperatures.

Chilling requirement: A cold period required by certain plants and plant parts to break physiological dormancy or rest. The chilling requirement is expressed in terms of the required number of hours at 7 degrees centigrade or less.

China Syndrome: The meltdown of a nuclear reactor.

Chipko movement: A local movement that began in India in the early 1980s and is opposed to governmental and other deforestation pro- grams.

Chlorofluorocarbons: Nontoxic chemicals used as coolants in refrigerators and air-conditioners, as propellants in aerosol cans, and as solvents. They are linked with ozone depletion and global warming.

Chronic: A health effect that takes a long time to manifest itself or that persists for some time.

Circle of poison: The use of pesticides banned in the developed countries on crops that are produced in less developed countries for export to developed countries.

Civil suit: A lawsuit centered on an individual seeking damages for injury or loss.

Class action suit: A lawsuit centered on a group seeking damages for injury or loss.

Class system: A system of social inequality based on the unequal distribution of economic resources.

Class: A person's ranking in a social hierachy; based on access to wealth and other scarce resources.

Clayey Soil: A soil is known as clayey which contain at least 30% clay particles and in most cases not less than 40%, usually it contain 45% clay, 30% silt and 25% sand.

Clean Air Act: An act passed by the U.S. Congress in 1963 to assure air safe enough to protect the public health.

Clean Water Act: An act passed by the U.S. Congress in 1972 to protect the nation's water resources, including regulation of pollution and sewage treatment.

Cleanup: An action taken to deal with the release of a hazardous substance that could affect the environment or human health.

Clear-cutting: A logging practice in which a majority of the trees in an area are cut.

Climate change: A change in climate caused by human activities or natural phenomena.

Climate: Aggregate of atmospheric condition over a long period of time.

Climate: The long-term average of weather con- ditions in an area.

Clone: A cultivar propagated by vegetative method is called a clone.

Cob: The large round mass of an ear of corn where kernels grow

Cognitive process: A mental process involved in human learning and reasoning.

Cold storage: A type of insulated storage utilizing mechanical refrigeration to maintain a stable, cold temperature for long-term storage.

Colluvial Soil: Are those which are form from the material transported by the force of gravity.

Colonialism: The practice of economic and po- litical domination of less developed countries by developed countries.

Combine: A machine used for harvesting grain

Command Area: Area which can be economically irrigated by an irrigation system.

Commercial Farming System: In this type of farming system, crops are raised on a commercial scale for marketing.

Commercial floriculture: Area of horticulture that includes the commercial production and distribution of cut flowers, flowering pot plants, foliage plants and bedding plants.

Commodities: Goods and services that can be bought and sold.

Commodity: An agricultural good.

Common law: Legal principles based on previous legal decisions.

Community: A group interacting at a specific time and place, as well as sharing a similar cultural background.

Compaction: The compression of air spaces in the soil by heavy machinery

Companion Crops: The two crops grown together are called companion crops. e.g. Berseem and barley.

Complete fertilizer: A fertilizer that contains nitrogen, phosphorus, and potassium.

Complete flower: Flower composed of a short axis or receptacle from which arise four sets of floral parts-sepals, petals, stamens, and pistils.

Compost: Decayed organic and animal matter used as fertilizer.

Compound leaf: A leaf with the blade divided into several leaflets or sections.

Compound umbel: A type of flower structure in which a series of simple umbels (see definition of umbel) arise from the same point on the main axis.

Condiment Crops: Crops which are grown and consumed as condiments e.g. coriander, mint.

Conidia: One celled asexual spores in certain fungi.

Conservation tillage: The practice of reducing or eliminating tillage operations and leaving crop residues on the soil to prevent erosion.

Conservationalists: Those wanting to preserve natural resources.

Consumptive Use of Water: Evapo-transpiration plus the water assimilated by various plant metabolic processes. As the water consumed in plant metabolism is very small, consumptive use and evapo-transpiration are considered almost equal.

Contagious diseases: Diseases that are transmitted by physical contact, e.g., tuberculosis and measles. They are the leading causes of death in many less developed countries.

Contaminant: Any substance that has an adverse effect on air, water, or soil.

Contingent valuation: Valuation of commodities not traded in markets, e.g., clean air, life expectancy, wildlife, and severed limbs.

Contour plowing: A soil conservation tech- nique in which cultivation follows the contours of the land.

Contraception: A birth control practice, such as use of condoms, intrauterine devices, or the pill.

Convenience food: Processed food often con- taining high levels of fat, sugar, and/ or salt; production and marketing are controlled by large transnational corporations.

Convention: A multilateral agreement between countries, usually a legal agreement on international environmental issues.

Conventional agriculture: Agricultural practices involving the use of chemical fertilizers, pesticides, and machinery.

Convergence thesis: The idea that different countries are becoming similar.

Core countries: Developed countries having the most technologically advanced, capital-intensive, and high-wage economies, e.g., the United States, Japan, and the western European countries.

Corms: Short, fleshy, underground stems with few nodes and very short spaces between nodes. Gladiolus and crocus are cormous plants.

Corn Ears: The part of a corn plant containing the corn cob, husk, and kernels

Corn Husk: The leaf like layer on the outside of corn ears, also known as a "shuck".

Cornucopian: A view that natural resources are unlimited.

Coronary heart disease: A disease affecting the heart and/ or coronary vein or artery. This is a leading cause of death in the developed countries and linked to overnutrition.

Corporation: A large private organization with multiple owners.

Corymb: Type of raceme flower cluster in which the pedicels of the lower flowers are longer than the pedicels of the upper flowers, resulting in a flattopped flower cluster.

Cost-benefit analysis: Evaluating projects, policies, and programs in terms of economic costs and benefits.

Cover crops: Plants used to hold the soil during the fallow season.

Cover Crops: The crops, which are planted to cover the ground and to reduce the soil erosion and nutrients losses by leaching. e.g grasses and rye.

Cowboy economy: An economy that behaves as if natural resources are infinite in supply and nature can absorb all wastes.

Cradle: A tool used for gathering a crop once it is finished growing

Crassulation Acid Metabolism (CAM) Plants: CAM plants fix CO_2 in four carbonic acid as do the C4 plants e.g pine apple.

Criteria pollutants: Substances that result in the most air pollution: carbon monoxide, sulfur dioxide, particulates, hydrocarbons, nitrogen oxides, ozone, and lead.

Critical Period of Competition: During the crop period there is a certain time when crop plants are most sensitive to competition by weeds, this time is known as the critical period of competition.

Critical Threshold Level (CTL): A weed, insect pest density capable of causing significant damage to crop is termed as critical threshold level.

Crop rotation: An agricultural method in which two or more crops are rotated from year to year to reduce nutrient depletion of the soil and reliance on pesticides.

Crop Rotation: Is the strategy of raising crops from a piece of land in such an order or succession that the fertility of land suffers minimally and the farmer's profits are not reduced.

Crop Water Requirement: The amount of water required to raise a crop to maturity with in a given period of time.

Crop: A crop is a community of plants grown under field condition for its economic value.

Cropping Intensity: The term cropping intensity refers to the ratio of actual cultivated area to total farm area over a year.

Cropping Pattern: It is a general cropping system followed or practiced by the farmers in an ecological zone.

Cropping Scheme: Allocation of an area to different crops being grown on a particular farm in a year.

Cross-pollination: The process in which pollen is transferred from an anther (the upper part of the stamen in which pollen is produced) of one flower to the stigma (the pollen-receiving site of the pistil) of a second flower of a different cultivar

Cultivar: A plant derived from a cultivated variety that has originated and persisted under cultivation, not necessarily referable to a botanical species, and of botanical or horticultural importance, requiring a name

Cultivate: To improve the land by plowing and fertilizing

Cultivator: Which only cut and stirr the soil

Culture: A commonly held set of beliefs, attitudes, and rules for behavior in a society

Cuticle: The waxy covering on leaves or fruit, which protects the tissue against excess moisture loss

Cuttings: Detached vegetative plant parts which when placed under conditions favorable for regeneration will develop into a complete plant with characteristics identical to the parent plant.

Cyme: A broad, more-or-less flat-topped determinate (having a fixed number) flower structure in which the central flowers bloom first

Day-neutral plant: A plant that will flower under any daylength

Debt crisis: A financial crisis faced by a number of less developed countries who borrowed extensively in the 1970s and early 1980s. Many of these countries have been unable to repay loans or keep up interest payments.

Debt-for-nature swap: A method begun in 1987 to help deal with the debt crisis by swapping debt for creation of nature preserves.

Deciduous: Trees or plants that shed their leaves every year.

Deep ecology: The environmental ethic maintaining that all species are of equal value and that humans have no right to reduce life except to satisfy basic needs. This often is identified as a biocentric worldview. It maintains that the anthropocentric worldview is the key cause of environmental problems. Proponents have been involved with radical environmental organizations such as Earth First, the Sea Shepherd Conservation Society, and People for the Ethical Treatment of Animals.

Define Necrosis: Death of organs of a plant, either as blight or death of tissue in localized areas, usually inside fruit and stems or die back or death of stems or branches.

Define Olericulture: Branch of horticulture which deals with cultivation of vegetables.

Deforestation: Forest loss; typically defined as a forest losing 40 percent or more of the trees.

Dehiscent: Type of dry fruit in which the carpel (a simple pistil) splits along definite seams at maturity. Determinate growth. Limited growth.

Delaney amendment (Delaney clause): An amendment to the Food, Drug, and Cosmetic Act that bans the use of all food additives that are carcinogenic.

Delta of Water: The depth of irrigation water required for the full crop period.

Democracy: A government in which the political power of elites is minimized and the political power of nonelites is maximized.

Demographic transition: The argument that as countries improve their standard of living, birth and death rates converge, and zero population growth is achieved.

Demography: The study of the size and composition of human populations, especially fertility, mortality, and migration patterns.

Dependence: The asymmetrical relations that characterize interaction patterns between countries occupying different positions in the world economy.

Dependency theory: The theory that dependent relations between nations foster positive development in the developed countries but distorted and constrained forms of development in the less developed countries.

Dependent development: A type of development characterized by economic growth but high income inequality and a repressive state. It is fostered in less developed countries by transnational corporations.

Depletion time: The actual time taken to deplete a nonrenewable resource.

Desalinization: The process of removing salt and related minerals from water for human uses.

Desert: A biome with limited precipitation (typically less than 10 inches per year) but different temperature ranges. Tropical deserts (Sahara) are hot year round, temperate deserts (Mojave) are hot in the summer and cool in the winter, and cold deserts (Gobi) are hot or warm in the summer and cold in the winter.

Desertification: The process by which land becomes desert through climatic change or human actions.

Determinate Plants: Those plant which initiate their reproductive stage after completing vegetative growth, e.g. wheat, barley.

Detritivore: An organism that feeds on dead organic matter.

Developed country: A country with a technologically advanced, capital-intensive, and high-wage economy.

Development: A value-laden notion referring to the extent to which a society is meeting the needs of its people. It typically is defined in economic terms, but encompasses other dimensions as well.

Dichlorodiphenyltrichloroethane: A chlorinated, hydrocarbon-based insecticide. It originally was considered safe and effective but proved to have negative human health and environmental consequences. It was banned in the U.S. in 1972.

Dicots: Dicots have two cotyledons and reticulate leaf venation.

Dieback: A sudden decline in the population of an area after the carrying capacity of the environment is reached.

Diet transition: A change in diet that is associatated with increased affluence. It typically moves from traditional grains (millet and sorghum) to rice and wheat and on to a combination of meats and grains.

Dioecious: Species that produce male and female flowers on separate plants.

Dioxins: Several chemicals created in the production of pesticides that have no industrial use but are hazardous to human health.

Direct action: A type of environmental protest practiced by environmental groups such as Greenpeace. It is characterized as nonviolent, passive resistance that forces the opponents to defend their position.

Direct regulation: A direct intervention in the market to regulate a hazard.

Dirty dozen: A group of 12 chemicals identified by the Pesticide Action Network as dangerous and posing a significant health risk.

Disbudding: The removal of vegetative or flower buds.

Discount rate: A rate used in cost-benefit analysis for discounting future values to the present. It typically refers to the value of something in the future when compared with current value.

Diseased plant: A plant that is abnormal because of a disease-causing organism or virus.

Diseases of civilization: Heart disease and cancer. Affluence increases fat consumption and causes a more sedentary and stressful lifestyle, which put people at risk for these diseases.

Disk: To prepare the soil for planting by cutting the soil with rotating metal disks

Diversified Farming: This is an expanded type of farming system in which varieties of crops are produced and many types of animals are reared.

Diversity: The number of species in an area.

Dobari Crops: A crop grown on residual moisture after the harvest of rice.

Domesticate: To tame and breed for human use

Dominant social paradigm: A western worldview maintaining that humans are superior to other creatures, the world provides unlimited resources for humans, and human history is characterized by substantial progress.

Dormancy: A state of suspended growth or the lack of outwardly visible activity caused by environmental or internal factors.

Dormancy: Seed dormancy is the state of inhibited germination of seeds with viable embryos in condition conducive to plant growth.

Dose: The level of exposure to a hazard.

Dose-response assessment: The determination of the relationship between dose or exposure and the intensity of the adverse effect.

Doubling time: The amount of time required for the population of a country to double.

Drainage: It is the removal of excess surface or ground water from the root zone of a crop by means of surface or sub-surface drains.

Drift net fishing: The use of huge nets that drift on the water to catch fish.

Drinking water: Water fit for human consumption.

Drip irrigation: Irigation using a tape or pipe with small holes that release water near the roots of plants and eliminate runoff.

Drip irrigation: The application of small quantities of water directly to the root zone through various types of delivery systems on a daily basis.

Driving forces: Social forces identified as the sources of environmental problems. Typical forces are population expansion, economic growth, political and economic institutions, technology, and cultural values.

Drought: The prolonged absence of natural precipitation.

Drupe: Type of fruit with a thin outer skinlike region, a center portion that is thick and fleshy, and an inner area that is hard and stony.

Dry farming: In which crops and livestock are raised on land which does not receive sufficient rainfall for water intensive crops and no irrigation facilities are available fall into this category.

Dual economy: An economy with a rich mod- ern sector and a poor traditional sector; often a problem for less developed countries.

Dust bowl: An area in the Great Plains that experienced drought and soil erosion in the late 1920s and 1930s.

Duty of water: The relationship between irrigation water flow and its commanded area where crops mature fully with that amount of water within a base period is called duty of water.

Earth Day: An event established in 1970 by Gaylord Nelson and held every year on April 22 to raise environmental awareness.

Earth Summit: The United Nations' Conference on Environment and Development, which took place at Rio de Janeiro, Brazil in June, 1992.

Earthing up: The operation of pulling up soil from the center of crop rows to the bottom of the plants, this helps in uprooting weeds and supporting to plants.

Eco: A prefix derived from the Greek word for house.

Ecocide: A planned effort to exterminate an ecosytem or part of it.

Ecofeminism: An environmental position maintaining that environmental problems can be traced to male-dominated institutions emphasizing competition, dominance, and individualism. Environmental problems will be solved only when male-dominated institutions are replaced with egalitarian, cooperative, and nonaggressive institutions.

Ecological economics: Nontradional economics that focus attention not only on allocation and distribution of resources but also on the larger ecosystem or environment. The economy is viewed as a subsystem of the larger and finite ecosystem. Attention centers on the flow of matter and energy from the environment as

Ecological limit: The carrying capacity of a given area.

Ecological niche: The role of a species in an ecosystem.

Ecology: The study of the environment and how living things interact with it

Ecology: The study of the relationship between all living organisms and the environment.

Economic depletion: The use of 80 percent of a nonrenewable resource.

Economic growth: Growth in the output of an economy. The gross national product (GNP) often is used to measure growth in the economy.

Economic system: A system of ownership, institutions, and allocative and distributive mechanisms of an economy.

Economic yield: That part of the crop for which crop is grown e.g. stem in sugarcane, root in the case of sugar beet.

Economic yield: The economically important part for which a particular crop is grown.

Economics: The study of the means by which humans produce, distribute, and consume goods and services.

Economy: A human system by which resources are produced, distributed, and consumed.

Ecosphere: Biosphere.

Ecosystem: A community of living and non-living things that interact by exchanging matter and energy

Ecosystem: An interacting system of a biological community (biota) and its nonliving environment (abiota).

Ecotage and ecoterrorism: Aggressive acts undertaken against corporations and other parties to protect the environment.

Effective rainfall: It is the part of the rainfall which forms a portion of the water requirement of a crop or which can be used by crop.

Efficiency: The amount of product produced per input unit of energy, labor, or material.

Effluent fee: A fee paid by a polluter to discharge noxious emissions into the air and water.

Effluent: A discharge of waste or other noxious material into the environment.

Electromagnetic pollution: Electronic and magnetic fields created by electrical circuits. They may represent human health risks.

Eleventh commandment: A commandment put forward by the population control movement that "Thou shall not transgress the carrying capacity of the environment."

Emigration: Population migration away from an area.

Emission: The release or discharge of gases or particulate matter into the air.

Endangered species: Organisms that are at risk of becoming extinct.

Endocarp: The inner area of the fruit wall. Entomology. The study of insects.

Energy conservation: Elimination of energy waste.

Energy efficiency: The amount of fuel needed to sustain a particular level of production or consumption. It typically is defined as annual primary energy consumption per dollar of gross domestic product.

Energy: The capacity to do work; usable power.

Entropy: The measure of the degree of disorder within a system; derived from thermodynamics.

Environment: All living and nonliving components by which an organism is surrounded and affected.

Environment: Physical surroundings; all that is around you

Environmental accounting: An effort to incorporate into measures of economic output the environmental consequences of economic production, such as soil depletion and air pollution.

Environmental change: A human-caused decline in the quantity or quality of a renewable resource.

Environmental currency: Monetary values that adequately reflect costs to the environment of human activities, by determining such fac- tors as energy flows.

Environmental degradation: Depletion or destruction of a potentially renewable resource such as soil.

Environmental ethics: Moral relations that hold between humans and the natural world.

Environmental impact assessment: The process of identifying and assessing environmental impacts associated with a project, policy, or program.

Environmental impact statement: A report identifying the likely environmental conse- quences of some project, policy, or program.

Environmental movement: A political move- ment to reduce resource depletion/ destruction and pollution.

Environmental pathway: An environmental path of hazard exposure; major pathways include the air, soil, and water.

Environmental Protection Agency: A U.S. regulatory agency established in 1970 to control pollution and conduct research on the environment.

Environmental refugees: Migrants from an area devastated by natural or technological hazards.

Environmental services: The restorative functions of nature, such as conversion of carbon dioxide to oxygen by plants.

Environmental sociology: A branch of sociology examining the interaction between humans and the natural environment.

Environmentalists: Individuals who attempt to curb resource depletion/ destruction and pollution.

Enzymes: Proteins that start a chemical reaction

Eolian soil: The soil which is formed by the material transported by winds from one place to another is called eolian soil.

Epidemiology: The study of environmental and other factors determining disease. Epidemiologic studies use human population data to examine the distribution and determinants of adverse health conditions.

Epidermis: The outer layer of cells of all parts of a young plant and of some parts of older plants, such as leaves and fruits. These cells are usually covered with a waxy substance, called the cuticle, that minimizes water loss.

Epigeal germination: It is derived from two words epi "above" and geas "earth". In this type of germination the cotyledons come out above the soil surface and generally turn green and act as first foliage leaves. e.g. bean, cotton.

Equity: A pattern of fairness.

Erosion: A process by which rock particles and soil are deposited in a new location through water or wind action.

Erosion: To wear away topsoil by water or wind and can be caused by intensive farming and overgrazing

Ethanol: A form of natural gas that can be produced from corn

Ethics: A code of behavior regarding what is appropriate and inappropriate.

European Community: A group of 12 countries: Belgium, Denmark, France, Greece, Ireland, Italy, Luxembourg, Netherlands, United Kingdom, Portugal, Spain, and Germany.

European Economic Community: An economic union established in 1958 to promote trade in western Europe that currently includes Belgium, Denmark, France, Germany, Great Britain, Greece, Ireland, Italy, Luxembourg, the Netherlands, Portugal, and Spain.

Evapotranspiration: It is the total loss of water due to its evaporation from land, plant and water surfaces and transpiration by vegetation per unit area per unit time.

Evolution: A process of biological or social change involving adaptation to an ecosystem.

Excrete: To get rid of waste, such as manure

Exhaustive crops: Crops, which feed heavily on the soil and deplete soil nutrients e.g. sorghum, tobacco.

Exocarp: The outer skinlike region of the fruit wall.

Expense: Cost or charge of money

Exponential growth: A growth pattern in which some entity doubles in size during a given time period.

Export of pollution: Transporting pollutants to another area or country.

Exposure assessment: Determination of the extent of contact between an organism and a hazard.

Extension Agency: An outreach arm of an agricultural university which provides educational programs on farming and does research

Extensive agriculture: Maximizing the amount of land used for agricultural production.

Extensive farming: In this type of farming large areas are used with minimum expenditure or attention to efficient use of other resources.

Externalities: Beneficial or harmful effects asso- ciated with production and consumption of a good that are not included in its market price.

Extinction: The disappearance of a species.

Factory farming: The mechanized, high-tech production of animals for human consumption. Growth hormones and antibiotics used to en- hance growth of animals may pose health risks to humans.

Fallow: A system in which land is left without a crop for an extended period to accumulate moisture and restore nutrients.

Fallow: Left wihtout tilling or sowing after plowing

Famine: Malnutrition and starvation resulting from a shortage of food.

FAO: See Food and Agriculture Organization.

Farm Bureau: A non-governmental political agency that works for farmers' rights

Fauna: The animal life of an area.

Feed Lots: A small area where cattle are confined and fed carefully mixed, high-concentrate feed to fatten them

Feed: A mixture or preparation used for feeding livestock

Feeder cattle: Cattle, ready to be finished for market, weighing 550-650 pounds or heavier

Feedlot: An area containing a high density of animals that are fattened by intensive feeding; high concentrations of wastes make them hazardous.

Fertilizer: A substance containing chemical elements needed for plant growth, mainly potassium, phosphorus, and nitrogen.

Fertilizer: Organic or inorganic nutrients that are added to the soil to help the growth of crops

Fiber crops: The crops, which are grown for their fiber and are used in making textiles, ropes. e.g. jute, sun hemp, cotton.

Field capacity: The amount of water retained by soil after drainage of saturated soil by gravitational force is called field capacity.

Field carrying capacity: It is the field capacity. It is also called normal moisture capacity.

Finances: Management of money affairs

First law of thermodynamics: A law stating that energy is neither created nor destroyed, but it does change form.

First World: Countries that were the first to industrialize.

Fission: The process of splitting atoms to release energy.

Fixing: In the nitrogen cycle, it is the process of nitrogen changing into a less mobile and more usable form by combining with hydrogen to make amonia

Flail: A wooden bar with a wooden handle used for removing grain or seeds from stalks

Fleshy fruits: Classification of fruits that includes the berry, drupe, and pome. They have a fruit wall that is soft and fleshy at maturity.

Flood plain: A low area along a river considered to be at risk of flooding.

Flora: The plant life of an area.

Floriculture: The study of growing, marketing, and arranging flowers and foliage plants.

Flower: A shoot of determinate (limited in number) growth with modified leaves that is supported by a short stem; the structure involved in the reproductive processes of plants that bear enclosed seeds in their fruits.

Fly ash: An air emission created by the incineration of solid wastes.

Food and Agriculture Organization: A United Nations Agency based in Rome and given the charge to improve efficiency of production and distribution of agricultural products.

Food chain: A hierachy of organisms, each feeding on the lower one.

Food gap: The difference between what people need to subsist and what is produced by farmers. This occurs in some low-income countries that are unable to produce basic foodstuffs.

Forage crops: Those crops, which are grazed by animals and harvested for green chop, hay, silage are classified as forage crops e.g. maize and sorghum.

Forcing: The manipulation of environmental factors which makes it possible to produce a marketable pot plant or cut flower out of season.

Forest: A biome with enough precipitation to support various tree species.

Fossil fuels: Mineral fuels that occur in rock formations. They include coal, oil, and natural gas and provide a majority of the energy used in

Foundation seed: Seed stock handled to most nearly maintain specific genetic identity and purity under supervised or approved production methods certified by the agency.

Free market: A market in which buyers and sellers are free to contract on whatever terms they like without government interference.

Friends of the Earth: An environmental advo- cacy group founded in 1970 by David Brower.

Frost pocket: A depression in the ground into which cold air drains but from which it cannot escape, thus causing it to be an area very subject to freeze injury.

Frugivore: An animal that lives on fruit. Frugivores are typically primates that cannot digest cellulose in leaves.

Fruit farming: In which orchards are planted and the objectives are to maximize fruit production, enhance quality and increase income.

Fruit: An expanded and ripened ovary with attached and subtending reproductive structures. Genes. Basic units of hereditary material that dictate the characteristics of individuals.

Fuelwood crisis: A shortage of wood for heat- ing and cooking purposes. This is a pressing problem in many less developed countries.

Functional Allelopathy: It is the case when the chemicals are toxic after being transformed by microorganism.

Fungicide: A pesticide that kills fungi.

Fungus: Parasitic or saprophytic plants lacking the green pigment chlorophyll and thus incapable of photosynthesis.

Fusion: The process of combining atomic nuclei in order to release energy.

Future Farmers of America (FFA): An educational organization for young adults who plan to have a career in agriculture

Gaia hypothesis: The hypothesis proposed by Lovelock and Margulis that living organisms on earth help regulate and stabilize the climate.

Garden crops: Vegetable crops, which are grown for their edible leaves, shoots, flowers, fruit and seed. e.g cabbage and okra.

Gene: The DNA code in the cells of all living things; they determine physical characteristics such as fur color

General Agreement on Tariffs and Trade: A multilateral trade agreement regulating the world trade system.

Genetic diversity: The genetic variability within a group of related organisms.

Genetic engineering: The manipulation of genetic material for economic purposes.

Geographic Information System: A computer software mapping system that provides a basis for manipulating, analyzing, and displaying spatial information.

Geosphere: The nonliving portion of the earth, excluding the atmosphere, hydrosphere, and biosphere.

Geothermal energy: Steam and hot water (created by the earth's molted core) used to produce electricity and considered to be a renewable energy.

Germination: Is the emergence and development from the seed embryo of those essential structures which, for the kind of seed provided, indicate the ability to produce a normal plant under favourable conditions.

Germination: The initiation of active growth by the embryo, resulting in the rupture of seed coverings and the emergence of a new seedling plant capable of independent existence.

Ghost acres: The land on which cheap food-stuffs and agricultural products are produced by less developed countries for export to the industrialized countries.

Global 2000: A report on global environmental trends that was commissioned by President Carter in 1977 and was published in 1980. The general conclusion of the report was that the world in 2000 would be more crowded, more polluted, and more fragile if global trends continued.

Global commons: Natural systems and resources that do not belong to any one country, e.g., the atmosphere and the oceans.

Global Environmental Monitoring System: A global effort to monitor the earth's environment; established in 1972 and operated by the United Nations Environment Programme.

Global positioning system: A U.S. Department of Defense satellite network used as a surveying and navigational aid and available to the public.

Global warming: An increase in the earth's surface temperature caused by the heat trapped in the earth's atmosphere by human-created gases.

Grade: Classification system of food quality

Grafting: The joining of two separate structures, such as a root and stem or two stems, so that by tissue regeneration they form a union and grow as one plant

Grange: An organization of farmers that provides support and plans social functions

Granivore: An animal living on a diet of grain or seeds.

Grass: A type of plant with jointed stems, slender flat leaves and spike like flowers such as corn and wheat

Grassland farming: These systems are mainly concerned with growing grasses for consumption by livestock kept for milk or meat production.

Grassland: A biome with a moderate level of precipitation and vegetation dominated by grasses.

Gravitational potential: It is produced by gravitational forces operating on soil water.

Gravitational water: Is the water in excess of hygroscopic and capillary water that percolates through the soil under the action of gravity if favourable conditions for water drainage are provided.

Great Plains: A grassland area in the western U.S., extending from North Dakota south to Texas and from the Rocky Mountains east to western Minnesota and Missouri.

Green consumerism: Purchasing environmentally sound products.

Green manure crops: Some crops are grown and ploughed in the soil in green form in order to improve soil fertility e.g. Berseem, Guara, Dhaincha etc.

Green politics: Politics centering on issues surrounding environmental problems.

Green revolution: The effort organized by the United Nations in the 1960s to increase world food production by introducing high-yield varieties of rice, wheat, and maize and new techniques, including irrigation and use of pesticides.

Green tax: A tax on activities that pollute, deplete, or degrade the environment.

Greenhouse effect: A rise in the earth's surface temperature caused by heat radiated by the sun that becomes trapped by greenhouse gases.

Greenhouse gases: Trace gases that contribute to the greenhouse effect, mainly carbon dioxide, methane, nitrous oxide, and chlorofluoro-carbons.

Greenpeace: An environmental organization emphasizing sea mammal protection and elimination of toxic pollution.

Greens: Political parties with an environmental stance.

Grit: The inside of a corn kernel exposed after the outer covering, or hull, is removed.

Gross domestic product: The value of goods and services produced in the country.

Gross national product: The value of goods and services produced in a country plus remittances received from abroad.

Gross primary production: The total energy produced by photosynthesis and stored in a given biotic community.

Groundwater: The water stored underground in rock and soil.

Groundwater: Water held underground that has seeped through soil layers and bedrock

Group of 10: The 10 most influential environmental organizations in the United States.

Group of 15: A group of 15 less developed countries that meet to discuss issues of the Third World.

Group of 7: Seven of the largest industrial countries, including Canada, France, Germany, Italy, Japan, the United Kingdom, and the United States.

Group of 77: A group of less developed countries in the Third World and Eastern Europe, originally including 77 countries but later expanded to include 129.

Growing season: The period from the last spring freeze until the first freeze in the fall. In the United States this ranges from about 100 to 365 days.

Growth mania: A belief that "bigger is always better".

Growth rate: The annual percentage increase in the gross national product

Habitat loss: The loss of natural habitat, often by human actions.

Habitat: The place where organisms live.

Hamburger connection: The clearing of land in Central and South America to produce cattle whose meat is exported to U.S. fast-food restaurants.

Hard pan: A hard semi impervious layer usually developed due to continuous ploughing to a depth of about 15 cm, with cultivators, or with continuous deposition of salts due to soil or surface irrigation water.

Hardening: The result of many changes that occur in a plant as it develops resistance to adverse conditions, especially cold.

Hardpan: A layer in the soil that cannot be penetrated and which restricts root penetration as well as movement of air and water.

Hardwood cuttings: Cuttings made from woody deciduous species and narrow-leaved evergreen species, such as grape and hemlocks.

Hardwood: The wood of broad-leaved, flowering trees, e.g., oak, mahogany, and walnut.

Harvest index: It is the ratio of economic yield to biological yield. It is expressed in percentage in H.I. = Economic yield x 100 / Biological yield

Harvest: To gather a crop when it is finished growing

Hazard identification: The process of determining that a technology, activity, or substance causes adverse health, safety, or environmental consequences.

Hazard: A technology, activity, or substance that has adverse effects on the environment or human health.

Hazardous waste: Waste possessing chemical, physical, or biological characteristics that represent a threat to either the environment or human health.

Heading back: Type of pruning cut where the end portion of the shoot is removed but the portion at the base is not.

Hegemonic power: The power exercised by states on an international basis because of their economic and military status.

Herbaceous cuttings: Cuttings made from herbaceous plants (see definition below), such as chrysanthemums, coleus, and geraniums.

Herbaceous perennials: Plants with soft, succulent stems whose tops are killed back by frost in many temperate and colder climates, but whose roots and crowns remain alive and send out top growth when favorable growing conditions return.

Herbaceous: Plants with soft and easily vulnerable body parts.

Herbarium: A collection of plant specimens that have been taxonomically classified, pressed, dried, and mounted on a sheet of herbarium paper.

Herbicide: A chemical used to control weeds and unwanted plants.

Herbicide: A material that will kill plants. Herbicides may kill essentially all plants or be quite selective in their activity.

Herbicide: A type of pesticide that kills weeds

Herbivore: An organism that obtains energy from plant consumption. Granivores and frugivores are special types of herbivores.

Herbs: Are plants of small to medium height and canopy.

High-pressure sprayer: A sprayer using a high-pressure pump to force the spray through nozzles for both creating a fine spray and delivery to the plant.

Hock: Back leg of cattle

Horticulture: The intensive cultivation of plants.

Human ecology: An area of inquiry concerned with the relationship of humans and their activities with the physical environment.

Human exemptionalism paradigm: A set of assumptions underlying the belief that humans transcend the environment in which they live. Specifically, given their unique cultural heritage, humans are different from all other species on earth, social and cultural factors determine human society, the larger environment is irrelevant, and all problems are soluble by human ingenuity and technology.

Humus: A dark-colored organic material that is the product of organic matter breakdown by both microorganisms and chemical reactions.

Humus: Sticky, brown part of the soil that comes from dead plants and animals and contains many nutrients

Hybrid: An offspring of two animals or plants that are of different breeds, varieties or species

Hybrid: In its simplest form, a first-generation cross between two genetically diverse parents. Imperfect flower. Flower lacking either the stamen or the pistil.

Hydrocarbons: Chemicals consisting of hydro- gen and oxygen and contributing to air pollution.

Hydroelectric power: Electricity created by movement of water.

Hydrophyte: Plant which grows in water, or which loves water.

Hydroponic: Grown in water without the use of soil

Hydrosphere: The water portion of the planet. Approximately 75 percent of the earth's surface is covered by water.

Hygroscopic water: Water attached to soil particles by loose chemical bonds and does not move by the action of gravity or capillary force.

Ice Ages: Periods when ice sheets moved from the polar cap and covered areas of North America, Europe, and Asia.

Ideotype: An ideal plant type developed through breeding.

Immigration: Population migration into another area.

Immiserization: An economic decline among poor people, making it difficult for them to meet their basic needs for food, health, and shelter.

Imperialism: A system in which one country uses the resources of a less powerful country for its own benefit, typically involving economic and political control.

Incineration: Burning wastes for purposes of disposal.

Income: A gain in money usually as a result of business or labor

Income: A gain or benefit expressed in money or in goods and services over a specified time period.

Incomplete flower: Flower lacking one or more of the fours sets of floral parts.

Indehiscent: Type of dry fruit in which the fruit wall does not split at any certain point or seam at maturity. Indeterminate growth. Growth that is potentially limitless.

Indeterminate plants: In these plants, the vegetative and reproductive stages continue simultaneously e.g. okra, tomato. Soybean is the only crop, which has determinate and indeterminate as well as semi-determinate growing types.

Index of sustainable economic welfare: An alternative measure for assessing the strength of an economy and human well-being. Unlike gross national product and similar indicators, it includes measures of environmental degradation.

Indicator species: A species that can indicate whether an ecosystem is being degraded.

Industrial economy: A production system based on machines to produce things of value.

Industrialization: A stage in societal development when resources are shifted from agriculture to manufacturing.

Infant mortality rate: The number of infant deaths (children between the ages of 0-12 months) per 1,000 live births.

Infectious diseases: See contagious diseases.

Inflorescence: The arrangement of the flowers on the floral axis-a flower cluster.

Informal economy: Business outside of the rec- ognized sector of a country's economy.

Inoculant: The bacteria containing material used to introduce N-fixing Rhizobium bacteria into soil.

Inputs: The amount of energy and money put into a farm in order to make a product

Insecticide: A chemical substance used to kill insects.

Insecticide: A type of pesticide that kills insects

Integrated pest management: A management effort combining biological and chemical controls to reduce reliance on synthetic pesticides.

Integrated weed management: The concept of IWM involves the planned use of all possible direct and indirect measures rather than relying on a single method to combat weeds.

Intensive agriculture: A system to maximize output of land through use of chemicals and machinery.

Intensive grazing: The practice of rotating livestock between pastures to reduce overgrazing

Inter cropping: Growing of two or more crops together on the same field, where one crop (main crop) planted in rows first and then another crop (intercrop) is planted in between the rows. These crops remain in association for a shorter time. These crops may or may not be planted and harvested at the same time e.g in Sugar cane planting of onion, garlic etc.

Interception: When the drops of rainfall or precipitation are intercepted by plant leaves it is called interception.

Intergenerational equity: A norm that calls for considering the interests of future generations when dealing with natural resources and a healthy environment.

Internal cost: The direct cost associated with a product or service that is paid by the producer and/ or consumer.

International regime: An international agree- ment between nations.

International Union for Conservation of Nature and Natural Resources: A nongovernmental organization founded in 1948 and concerned with threats to the quality of the natural environment, especially wilderness areas and endangered species.

Interstock: Intermediate stem piece that is grafted between the upper and lower pars of the graft.

Inventory: A list of current goods that you have or own.

Irrigated soil: The soil, which receives irrigation water from well and tube wells, are known as irrigated or chahi soil.

Irrigation efficiency: It is a term used to indicate how efficiently the available water supply is being used for crop production.

Irrigation scheduling: It refers to the number of irrigations for a crop and their timing.

Irrigation water requirement: The quantity of water required for successful crop production exclusive of precipitation, ground water and other natural resources.

Irrigation: Irrigation is the artificial application of water to soil or crop plants to assist crop production.

Kera: Manual sowing of seeds in lines in furrows.

Kernel: The seed of a grain plant

Kharif crops: Those crops, which are planted in the summer month from the March to July and harvested in autumn and winter, are called Kharif crops e.g. rice and cotton.

Lacustrine soil: When the material transported by streams and river, if deposited in lakes the soils are called Lacustrine soil.

Landscape design: The profession concerned with the planning and planting of outdoor space to secure the most desirable relationship between land forms, architecture, and plans to best meet human needs for function and beauty.

Latitude: The angular distance north or south of earth equator.

layering: A vegetative method of propagation that produces new individuals by producing adventitious roots (roots growing from unusual places) before the new plant is separated from the parent plant.

Leaching: The downward movement of nutrients or salts through the soil profile in soil water. Leaching accounts for nutrient losses but can also be beneficial in ridding a soil of excess salts.

Leaf arrangement: The arrangement of leaves on the stem.

Leaf-bud cuttings: Cuttings consisting of a leaf blade, leaf stalk, and a short piece of the stem. Leaf cuttings. Entire leaves with or without the leaf stalks.

Leaflets: Individual parts of a compound leaf.

Leaves: Vegetative plant parts that are lateral outgrowths of stems that have developed special structures for photosynthesis.

Legume: A simple, dry, dehiscent (see definition of dehiscent) fruit with one locale that splits along two seams.

Legume: Any plant that grows seeds in a pod such as peas and beans

Lime: Ground limestone that is used to raise soil pH.

Line: A cultivar propagated by seed is called a line.

Livestock and poultry farming: This category includes farming system in which various kinds of livestock are reared for meat, milk, wool and eggs.

Livestock: Any animals raised on the farm

Loam: A type of soil consisting of sand, clay, and organic matter. Locales. The cavities of the ovary of the pistil of a flower.

Loamy soil: The soil is more or less than midway clay and sandy soil and ideal loam soil may defined as a mixture of sand, silt and clay particles which shows the properties of sandy, silt and clayey in equal proportion.

Lodging: The bending or breaking over of a plant before harvesting.

Long day plants: Plants which change vegetative to reproductive stage by producing flowers and fruits, when the days become longer. e.g. Carrot, Radish etc.

Long-day (short-night) plant: A plant that requires a day longer than its critical daylength to flower.

Malling series: A group of apple rootstocks that originated at the East Malling Research Station in England.

Malling-Merton series: A group of apple rootstocks that were the result of the crossing the Malling series with the 'Northern Spy' to add resistance to woolly aphids.

Manure: Animal waste from stables or barnyards

Market gardening: The growing of an assortment of vegetables for local or roadside markets. Mesocarp. The center portion of the fruit wall.

Market: Where products are sold and exchanged

Matric potential: It is produced by capillary and surface forces.

Matter: The material which makes up something

Maximum potential soil moisture deficit: Is the greatest value of potential soil moisture deficit attained during the growth of a crop.

Micronutrient: An essential nutrient that is needed in small amounts; also called a trace or minor element. Monoecious. Plants that produce male and female flowers on the same plant.

Microorganisms: Tiny living things that can only be seen with a microscope

Mill: A machine used to grind grain for food

Monocots: These plants have one cotyledon and parallel leaf venation.

Monocropping: Raising of only one crop in a year when there is seasonal supply of water is called Monocropping, monoculture or specialized farming.

Monoculture: Planting the same crop in a field year after year with no crop rotation

Mound or stool layering: A layering procedure in which the new shoots develop in spring and soil is mounded around their bases, excluding light and enhancing root formation.

Muck soil: If the quantity of organic matter exceeds 20% but less than 50% are called muck soil.

Mulch: A material applied to the surface of a soil for a variety of purposes such as conservation of moisture, stabilization of soil temperature, and suppression of weed growth.

Mulch: Any material or practice which is used to check the loss of water by evaporation is called mulch.

Multiple cropping: It is growing of two or more crops in a year from the same piece of land

Multiple cropping: Raisingof two or more crops in the same field or in a year is called diversified multiple cropping or simple multiple cropping.

Narcotic or drug crops: This category includes those crops, which have some narcotic and drug value e.g. poppy, tobacco.

Natural erosion: The erosion of the soil under natural condition.

Natural soil: Strictly speaking a soil having pH of 7, in practice a soil having PH 6.6 and 7.3.

Net plot: Area from which yield and other characteristics are measure. It is also known as the net area of the plot.

No till: A method of growing grain when the farmer does not plow the field before planting

No tillage crop (zero tillage crop): Crop grown with out any tillage to prepare seed bed or row.

Node: A slightly enlarge portion of stem where leaves and bud arises and where branches originate.

Nodule: A swelling on the root of a legume that contains bacteria that fix nitrogen from the air

Nucleus seed: Seeds obtained from selected individual plants of a particular variety which needs to be purified and multiplied in such a way as to maintain its genetic purity.

Nursery grower: A person who produces or distributes ornamental plants.

Nut cycle: The regeneration/cycling of nutrients.

Nut: An indehiscent dry fruit similar to the achene except that the fruit wall is hard throughout. Offsets. Miniature bulbs grown to full size.

Nutrients budget: A quantitative data of the major nut flowing to retained within the discharge from the system.

Nutrients: Nitrogen, phosphorous, potassium, and minerals that plants need to grow

Nutrients: The food for microbial and plant life mainly composed of nitrogen and phosphorous but also of potassium, Mg, Fe, Ca, Co, Cu, Zn and others elements.

Offshoots: Short, horizontal stems that occur in whorls or near whorls at the crown of stems.

Oil seed crops: These are the crops, which are grown for the purpose of extracting oil from their seed e.g. mustard and groundnut.

Olericulture: The study of vegetable production.

On-farm water management: The planned use of irrigation water at the farm level or more efficient utilization in agriculture.

Opposite leaves: Leaves arising from opposite sides of the same node.

Organic farming: Producing foods without the use of laboratory made fertilizers, growth subtances, or pesticides

Organic matter: Materials rich in carbon of either plant or animal origin, which exist in all stages of decomposition of soils.

Organic matter: The dead plants, animals and manure converted by earthworms and bacteria into humus

Osmotic potential: It is also called solute potential. It is produced by various solutes in soil water.

Ovary: Part of the pistil that contains one or many small bodies known as ovules. Ovule. The immature seed in the ovary.

Palmate leaf: A compound leaf with all the leaflets arising from one point at the end of the leaf stalk

Palmately veined leaves: Leaves with netted veins having several large veins radiating into the blade from the leaf stalk at the level where the leaf stalk and blade join.

Panicle: Type of flower cluster with either a cluster of racemes or corymbs (see definitions), distinctly branched.

Parallel venation: Leaves with large veins that are essentially parallel to one another and are not connected by lateral veins.

Pastures: Land or a plot of land used for the grazing of animals

Peat soil: If the quantity of organic matter is more than 50% is called peat soil.

Peduncle: The short stem of the flower cluster.

Pepos: Berries that have a hard rind around the fruit.

Percolation: Downward movement of water with in the soil profile.

Perennials: Plants that do not die after flowering but live from year to year.

Perfect flower: A flower that has both a pistil (or pistils) and stamens but may lack sepals, petals, or both. Perianth. Structure consisting of the sepals and petals.

Pericarp: The fruit wall, consisting of three distinct layers; the exocarp, the mesocarp, and the endocarp.

Permanent wilting percentage: The soil water content at which plants can no longer extract sufficient water from the soil for their growth is called permanent wilting percentage.

Pest management: The control of a pest or group of pests by a broad range of techniques from biological control to pesticides. The goal is to keep damage below economic levels without completely eliminating the pest.

Pesticide: A substance that kills any pest, including insects, fungi, and weeds

Petals: Structures collectively making the corolla, which protect the inner reproductive structures and often attract insects by either their color or their nectar and thus facilitate pollination.

Petiole: The leaf stalk.

pH: A measure of acidity or alkalinity, expressed as the negative log of the hydrogen ion concentration. A pH of 7 is neutral; less than 7 is acidic more than 7 is basic

pH: A value that indicates the acidity of the soil

Photosynthesis: The process by which green plants use light energy from the sun to produce sugar from water and the air

Pinching: Breaking off the uppermost growing point, thus allowing the willary buds to start to grow. This promotes outward growth rather than upward growth.

Pinnate leaf: A compound leaf with the leaflets arranged along both sides of the midrib.

Pinnately veined leaves: Leaves with netted veins where secondary veins extend laterally from a single midrib.

Pistil: The female reproductive organ, consisting of the stigma, style, and ovary. Plant growth. A permanent increase in volume, dry weight, or both

Plant development: Plant development is the progress of plant from germination to maturity through a series of stages.

Plant growth: It is the increase in the dry weight of a plant over time mainly as a consequence of photosynthesis.

Plant pathogen: A microorganism that causes a plant disease.

Plant propagation: Increase in numbers or perpetuation of a species by reproduction.

Plough pan: A dense, compacted layer about 5 to 7 cm thick formed beneath the surface soil by repeated ploughing in the same path.

Plough: Its function is to cut, stirr, invert, and pulverize the soil.

Plow: To work the soil by turning over the top layer; the machine used to turn the top layer of soil

Pollinate: To fertilize by transferring pollen from the anther to the stigma of a flower

Pome: Type of fruit in which the portions produced by the pericarp are enclosed within fleshy parts that are derived from parts of the flower other than the ovary. An apple is a pome.

Pomology: The study of fruit production.

Pore space: It is a space between soil particles occupied by air and water; it is largely controlled by the texture of soil.

Pork bellies: Meat from the belly area of a pig; used for bacon

Port: A town or city where ships are loaded with products to be shipped overseas

Pot-bound: Having restricted root growth and a circular pattern of root growth caused by a too-small container.

Potential Evapotranspiration: Is defined as the amount of evaporation occurring from an extensive area of a short, green growing crop completely covering the ground and well supplied with water.

Potential soil moisture deficit: It is the difference between a crop potential evapotranspiration and the amount of rainfall received by a crop plus the quantity of water delivered to it in irrigation.

Pre-basic seed: It is the progeny of nucleus seed, and is handled so as to maintain specific genetic purity and identity as completely as possible.

Pressure potential: It is produced by actual hydrostatic pressure.

Pruning: Removal of plant parts such as buds, developed shoots, and roots to maintain a desirable form by controlling the direction and amount of growth.

Puddling: Ploughing in standing water to create a shallow hard pan at a 10 to 15 cm depth, which helps to increase water-holding capacity and reduce moisture losses by percolation.

Pulses or grain legumes: The crops belonging to Leguminoseae family are grown for their edible seed e.g chick pea, lobia.

Rabi crops: These crops are planted in winter from October to December and harvested in summer from March to May e.g. wheat,

Raceme: Type of flower cluster in which stalked flowers are on pedicels approximately equal in length on a single floral axis.

Readily available water: The portion of the available water that is most easily extracted by a plant is called readily available water.

Relay crops: A relay crops is one which is planted as a second crop after the first crop has reached its reproductive stage of growth but before it is ready for harvest. e.g. planting of sugar cane in sugar beet.

Relief: When rising ground or mountains running at right angle to the prevailing wind.

Rendering plant: A place where lard, tallow, and oil are extracted from animal parts

Rest: A state of suspended growth or outwardly visible activity due to internal physiological factors. Also referred to as physiological dormancy.

Rhizome: Horizontal stems that grow underground. They serve as storage organs and a means of vegetative reproduction.

Root and tuber crops: These are vegetable crops grown for their under ground parts like roots, bulbs, rhizomes, corms and stem tubers e.g. carrot and onion.

Root cuttings: Cuttings 5 to 15 cm long made from root sections in the fall or winter.

Root: Vegetative plant part that anchors the plant, absorbs water and minerals in solution, and often stores food.

Rotation: The changing of the specific fields used for one crop year to year

Row Dividers: The large points on the end of a combine used to pick-up corn

Rumber: The process in which planker or leveler is used to conserve moisture at watter condition before the preparation of land after rauni.

Rumen: The first large compartment of the stomach of a bovine; its bacteria and protazoa break down cellulose

Run off: When water flows out the field by breaking the bunds of the field or flows to the sloppy areas from the high level is called run off..

Runner: A slender stolon (see definition) with elongated internodes. These root at the nodes that touch the ground. Example: strawberries.

Rust: A type of fungal disease that requires two hosts. For example, White Pine Blister Rust alternates between white pine and Ribes sp. Gives a rusty appearance to the plant.

Samara: An indehiscent dry fruit with either one or two seeds, in which pericarp bears a flattened wing-like outgrowth. The "helicopters" of the maple trees are double-samaras.

Sandy soil: The soil which contains 2.00 to 0.2 mm diameter soil particles, it contain 85% sand, and >15%, silt and clay. These soils are poor in plant material.

Saturation capacity: This term refers to the amount of water present in the soil when it is completely saturated with water.

Scales: Modified leaves that protect structures.

Scaling: A type of propagation that separates bulb scales from the mother bulb to induce the formation of bulblets.

Scarification: Any physical or chemical treatment that makes the seed coat permeable is known as scarification.

Scarification: The chemical or physical treatment given to some seeds to break or weaken the seed coat sufficiently for germination to occur.

Scion (cion): The upper part of the union of a graft.

Scythe: A blade with a long handle used to cut grass, grain, and other crops

Seed certification: Is the process to secure, maintain and make available high quality seed and propagating materials of superior crop varieties, so grown and distributed as to ensure desirable standards of genetic identity, physical purity and quality attributes.

Seed: Plant embryo with associated stored food encased in a protective seed coat.

Seedling: A young plant that has grown from a seed

Seeling: Mechanical weeding process carried out by ploughing with bullocks or tractors.

Seepage: Losses through the beds and banks of canals and water channels take place mainly by unlined canals.

Self-pollination: The process by which pollen is transferred from the pollen producing section of the plant to the pollen receiving part of the plant of the same flower or another flower of the same cultivar.

Semihardwood cuttings: Cuttings made from woody, broad-leaved evergreen species such as ligustrum and holly.

Sepals: Structures that usually form the outermost whorl of the flower; collectively, the calyx.

Separation: The use of bulbs and corms in propagation that utilizes the naturally detachable parts.

Serrate leaf: A leaf with serrations or "teeth" along the edge of the leaf.

Sessile: Without a petiole (as in some leaves), or without a pedicel (as in some flowers and fruits).

Sett: A piece of seed cane with two-four buds (eyes).

Sexual reproduction: The reproduction of plants through a sexual process involving cell division.

Shear: The act of cutting hair or wool

Shock: A pile of grain that is set up like a cone

Shoot: Stem, 1 year old or less, that possesses leaves.

Short day plants: Plants which changes from vegetative to reproductive stage and produces flowers and fruits, when the days become shorter.

Shrubs: Are bushy plants with medium to tall height and canopy.

Silage crops: Those crops, which are harvested when still green and succulent and are fed directly to animals with out curing, are called silage crops. e.g. Berseem, Shaftal, maize.

Silage: A mixture of raw materials such as field corn, sorghum, grass, or clover that is converted into winter feed for livestock

Simple bud: Bud containing either leaf or flower primordia, but not both.

Simple layering: A method similar to tip layering, except that the stem behind the end of the branch is covered with soil and the tip remains above ground.

Simple leaves: Leaf blades consisting of one unit.

Sinus: The indentation of a leaf. Sinuses alternate with lobes.

Slaughterhouse: A place where animals marketed for meat are killed humanely

Slow-release fertilizer: A fertilizer that is made by coating the particles with a wax or other insoluble or very slowly soluble material to provide a predictable, slow release of the encapsulated materials.

Softwood cuttings: Cuttings taken from soft, succulent, new spring growth of deciduous or evergreen species of woody plants.

Soil Conservation: Careful preservation or protection of soil

Soil management: The practices used in treating a soil, which may include various types of tillage and production systems.

Soil structure: The arrangement of individual soil particles.

Soil: A natural body developed from variable mixtures of broken and weathered minerals and decaying organic matter which covers the earth crust in thin layers and supplies proper amount of nutrients and air water and mechanical support to plants.

Soil: The outer, weather layer of the earth's crust that has the potential to support plant life. Soil is made up of inorganic particles, organic matter, microorganisms, water, and air.

Species: A group of living things that share common biological characteristics

Spike: Flower cluster of an indeterminate (unlimited number) raceme of sessile flowers attached to the floral axis with the oldest flowers at the base.

Spikelet: Diminutive type of spike inflorescence found in grasses.

Spines: Sharp-pointed woody structures, usually modified from a leaf or part of a leaf.

Stamen: Part of the flower consisting of the anther, in which pollen is produced, and a slender filament that holds the anther in a position favorable for pollen dispersal. Male reproductive organ.

Staminate flower: Flower in which only the stamens (male reproductive parts) are present; there are no pistils (female reproductive parts).

Steer: A bull that has been castrated for better meat production

Stewardship: An individual's responsibility to exercise care over possessions entrusted to him or her

Stigma: The pollen-receiving site of the pistil.

Stipules: Leaflike appendages often found on either side of the base of the petiole; they may subtend the leaf, but they are not actual parts of the leaf.

Stock (rootstock): The lower part of a graft.

Stolon: An aboveground stem that reclines or becomes prostrate and may form roots at the nodes that may come into contact with the ground.

Stratification: Is the practice of exposing imbibed seeds to cool temperature conditions for a few days prior to germination in order to break their dormancy.

Stratification: The storing of seeds at low temperatures under moist conditions to break physiological dormancy or rest.

Sub-division: A piece of land to be divided into smaller lots, typically for housing

Subsistence farming: In which basic necessities like food, clothing, and shelter are produced for the family to live on.

Succulent leaves: Water storage structures characteristic of plants in and and semiarid regions.

Sugar crops: These are the crops, which are grown for sugar purposes. e.g. sugar cane and sugar beet.

Taproot: Primary root that persists and maintains its dominance.

Technology: Instruments, tools or inventions developed through research to increase efficiency

Temperate fruit: A fruit plant that requires a cool period and is deciduous. Examples are apple, pear, and peach.

Temporary wilting percentage: Soil water content at which plants wilt during the hot windy part of the day but regain turgidity during the cooler part of the day is called TWP.

Tenant: A person who pays rent to live on someone else's land

Tendrils: Slender, twining leaf modifications used for support, as in the terminal leaflets of the grape.

Terminal buds: Large and vigorous buds at the tips of stems, responsible for terminal growth the world.

Thorn: A short, sharp-pointed branch.

Thresh: To beat out a grain or seed from a stalk by treading, rubbing or striking with a flail, or with a machine

Tillage: It is usually defined as the mechanical manipulation of the soil aimed at improving its physical condition.

Tiller: To put forth shoots other than the mother shoot from roots axis.

Tilth: It is the final condition of soil when all the tillage operations are carried.

Tilth: The physical condition of soil is called Tilth.

Tip layering: Layering in which rooting takes place near the tip of the current season's shoot, which naturally falls to the ground.

Tissue culture: The growing of masses of unorganized cells on agar or in liquid suspension. Useful for the rapid asexual multiplication of plants.

Topping: Topping in tobacco is the removal of the terminal bud with or without some of the small top leaves just before or after the appearance of the flower head.

Topworking: Grafting procedure by which branches of trees are changed to a more desirable cultivar.

Total Mixed Ration: A feed combination of hay, corn, barley, field grasses, cotton seed, and bakery or grocery by-products

Transpiration ratio: It is the ratio of the weight or volume of water transpired by the plant during its growth period to the weight of dry matter produced by the plant.

Transpiration: It is the process of loss of water from living plants.

Trickle irrigation: The application of small quantities of water directly to the root zone through various types of delivery systems on a daily basis.

Truck farming: It refers to the system in which the bulk of the out puts are produced and marketed.

True Allelopathy: When the allelochemical is toxic in their original form is called true Allelopathy.

True leaves: The first leaves to emerge after the germination of a seed.

Tubers: Enlarged underground stems serving as storage organs of starch or related materials. Potatoes are tubers.

Twig: Stem, 1 year old or less, without leaves.

U.S. Congress in 1973 to protect species in danger of becoming extinct.

Umbel: Type of flower cluster in which the pedicels arise from a common point and are about equal in length.

Urbanization: The growth of the city into rural areas

USDA: United States Department of Agriculture, a Federal agency involved in all phases of agriculture

Variegated: Leaves with green and white striped pattern instead of the typical all-green look.

Variety: In general the term variety has been used to refer to a group of similar plant with in a particular species that is distinguished by one or more then one character and given the name.

Vegetable: The edible portion of an herbaceous garden plant.

Vernalization and chilling: Many biennials and temperate annuals, as well as certain fruit trees, require exposure to cold temperature before they can flowers. This is known as Vernalization requirement for annuals and biennials and chilling requirement for fruit trees.

Vines: Are plants, which have tender stems, and requires some support for upward growth.

Water holding capacity: Ability of soil to retain water.

Water potential: Refers to the chemical potential of water.

Watter: Describes the condition when soil moisture level is suitable for cultivation.

Weather: It is a state of atmosphere at any time; it is combined effect of many things such as heat, cold.

Weed: A plant growing where it is not wanted.

Weed: any plant growing out of its proper place.

Weed: Any unwanted plant, especially those that crowd out more desirable plants

What are the compounds which act as buffer in soil? Clay organic matter and such compounds as carbonates and phosphates which make the soil to resist appreciable change in pH value.

What is Agronomy? Agronomy is the science of crop production. It incorporates the basic sciences (Biology, Chemistry, Physics, Geology, Microbiology, etc.) into an applied science which is the foundation for most agriculture.

What is arid? A term applied to land that is dry also means deficient in rainfall.

What is composite variety? A mixture of genotype from general sources maintained by normal pollination.

What is compost? A manure derived from decomposed plant remains usually made by fermentation, waste plant material under controlled conditions. Compost usually used in green houses to enrich the soil either dungs as surface.

What is contact herbicide? Which kill only those parts of plant with which they come into contact and used mostly to control annual weeds when seedling, they have little residual effect.

What is crop rotation? A definite succession of crops following one an other in a specific order.

What is drought resistance? Characteristics of plants which are suitable for cultivation in dry condition regardless of the inherent mechanism that provides resistance.

What is effective rainfall? Precipitation which falls during the growing period of the crop and is available to meet the Evapotranspiration requirements of crops .

What is Gibberellins? Plant growth stimulating chemicals which are able to induce a number of effects on plants.

What is harvest index? The ratio of grain weight to total plant weight in a cereal crop.

What is hybrid vigour? Qualities in a hybrid not present in either parent.

What is malathion? An Organophosphorus insecticide and acaricide available either alone or in mixture used in liquid or dust form.

What is meant by autotrophe? Organism able to manufacture their own food from inorganic materials. Using energy from outside sources. Most green plants are completely autotrophe.

What is meant by bolting? Formation of elongated stem or seed stalk, it is usually takes place during the second season of the growth in biennial plants.

What is meant by cropping intensity? Refers to the no. of crops which are raised during the year.

What is meant by F1 and F2 generation? Genetic terms for the off spring generations produced by a parental generation of plants or animals.

What is meant by layerage? Methods of plant propagation by asexual means in which a portion of stem, shoot or branch has covered with soil or some other medium in which root can develop , after which the rooted portion get detached from the parent plant.

What is meant by soiling crops? Crops harvested when green and succulent condition are fed to animals shortly after cutting, it is neither dried nor stored for future feeding.

What is organic farming? System of farming which avoid the use of artificial fertilizers, pesticides or herbicides, and concentrate on methods of crop rotation and the use of home growth feed, organic fertilizer.

What is pasteurization? Process of killing organisms in a product, commonly milk by heating to a controlled temperature.

What is pedology? The study of soil.

What is planker? Implement used to crush clods on land where a roller can be used consisting of number of fixed over lapping plates, showed with iron bars along the working edges which is pulled over the land.

What is seed certification? Refers to the system of maintaining the genetic purity and quality of seed.

What is seed dressing? The chemical treatment of seeds particularly cereals, with fungicides and some time insecticides to protect them against soil and seed borne disease and pest.

What is trench layering? An asexual reproductive method of plant propagation involving laying down the whole stem, the new shoots are thus forced to push their way through a layer of soil which prevents the bark from coloring and favors root formation.

What is truch gardening? Growing of crops like potato onion and cabbage on large scale for distinct market.

What is weathering? The process by which soil disintegrates and decomposes, eventually producing soil particles by exposure to the physical and chemical effects of atmospheric agents.

Whorled arrangement: Three or more leaves at a node.

Wilt: Loss of rigidity and drooping of plant parts caused by many different plant diseases. Generally, it involves an interference with water conduction inside the stems.

Wilting point: The point at which the water content of a soil reaches such a level that it is firmly held by soil and unavailable to plant roots, so that the plants wilt permanently and die.

Woody perennials: Plants with woody fiber.

Work ethic: Qualities of character believed to be promoted by work

Yellows: A type of disease, common among some flowers, that is caused by a virus or mycoplasma-tike organism. This disease causes stunting and a yellowish appearance.

Yield: The amount of a crop produced in a given time or from a given place

Zaid Kharif crops: These are the crops, which are planted in August to September and harvested in December to January e.g. toria.

Zaid rabi crops: These are crops, which are planted in February and harvested in May-June e.g. tobacco.